TRANSACTIONS OF THE
AMERICAN PHILOSOPHICAL SOCIETY
HELD AT PHILADELPHIA
FOR PROMOTING USEFUL KN

VOLUME 67, PART 4 · 19

The Invention of the Telescope

ALBERT VAN HELDEN
ASSOCIATE PROFESSOR OF HISTORY, RICE UNIVERSITY

THE AMERICAN PHILOSOPHICAL SOCIETY
INDEPENDENCE SQUARE: PHILADELPHIA
June, 1977

Library of Congress Catalog
Card Number 77-76425
International Standard Book Number 0-87169-674-6
US ISSN 0065-9746

FOREWORD

It is with genuine delight that one encounters nowadays such solid and substantial research into a subject of prime historical importance as Albert Van Helden's investigation of the origins of the telescope. Ours is an age of science and technology, based on precision instruments. The first such device to strengthen our feeble human senses in our irrepressible striving to comprehend the strange and elusive universe around us was the telescope. To whom should grateful mankind give thanks for the invention, or first invention (as our ancestors used to say) of this marvelous aid to sight? From its very inception the telescope has been wrapped in mystery. To wrest its secret from a confusing mass of conflicting statements and contradictory reports required, above all, tenacity. That indispensable quality is possessed in the highest degree by Albert Van Helden. In addition, he has benefited from a first-rate technical education, and also from a firm grasp of the requisite languages, principally Dutch (his native tongue), Latin, German, Italian, and French. He has here assembled the pertinent documents, translated them into English, and analyzed their implications. For thus sweeping away the cobwebs of obfuscation that have long veiled the origin of the telescope in a virtually impenetrable darkness, all serious students of the history of science and technology, particularly those with a love of precision instruments, will surely wish to join me in hailing Albert Van Helden as the latest and best investigator of this fascinating problem.

EDWARD ROSEN

PREFACE

This project began as a translation of Cornelis de Waard's *De uitvinding der verrekijkers* (The Hague, 1906). De Waard had uncovered many new documents bearing on the genesis of the telescope, but his work remained relatively inaccessible in its original Dutch. Only summaries and second-hand accounts of it found their way into other languages, and without the full weight of de Waard's copious new evidence his argument was not well received in the English-speaking world of historical scholarship. A translation of the book, therefore, seemed a useful project.

After I had become thoroughly familiar with the documents and with de Waard's argument based on them, however, it became apparent that intensive editing would be required to bring the historical apparatus up to date. I decided at this point that the profession as well as de Waard's memory would be better served by a collection and translation of all the relevant primary sources than by a translation of *De uitvinding der verrekijkers.* My great debt to Cornelis de Waard, however, will be apparent to all those who are familiar with his book.

The argument presented in the introductory essay here has evolved over a considerable period of time. Much of it was, so to speak, coaxed out of me by Derek J. de Solla Price, in conversations over a period of three years. I owe many of the ideas presented below to him. But Professor Price is in no way responsible for weaknesses and errors which the reader may find in my argument. I take comfort in the knowledge that, if the reader does not agree with my interpretation of the sources, he or she may nevertheless find the collection of documents convenient and useful.

My thanks go to A. Rupert Hall and Marie Boas Hall for their help with translations and for their patient reading of several drafts of this essay; to Stillman Drake for his thoughtful comments and suggestions; and to Edward Rosen for his careful reading of the final draft and his enthusiastic encouragement. My thanks also go to Jeanette Walthall for her patient typing of several drafts of this monograph.

A. V. H.

THE INVENTION OF THE TELESCOPE

Albert Van Helden

CONTENTS

INTRODUCTION

Among the scientific instruments which have played an important role in the growth of man's knowledge of the world around him, the telescope occupies a position of historic preeminence, rivaled only by the microscope (which was a natural outgrowth of the telescope).[1] In a real sense, the telescope can be considered the prototype of modern scientific instruments, and learned men in the seventeenth century—the first century of its existence—were acutely aware of its important role in the formation of a new astronomy. No wonder, then, that even during the early days of its existence a number of men became interested in the genesis of the instrument.

News of the existence of a device for seeing faraway things as though they were nearby spread through Europe beginning in the autumn of 1608, and less than two years later Galileo published his epoch-making *Sidereus nuncius,* in which he discussed the celestial discoveries made with the new instrument. In this work, Galileo mentioned how he had first become aware of the existence of the new *occhiale*: [2]

> About ten months ago a report reached my ears that a certain Fleming had constructed a spyglass by means of which visible objects, though very distant from the eye of the observer, were distinctly seen as if nearby. Of this truly remarkable effect several experiences were related, to which some persons gave credence while others denied them. A few days later the report was confirmed to me in a letter from a noble Frenchman at Paris, Jacques Badovere, which caused me to apply myself wholeheartedly to inquire into the means by which I might arrive at the invention of a similar instrument. This I did shortly afterwards, my basis being the theory of refraction.[3]

Although Galileo clearly indicated here, as well as in *Il saggiatore* of 1623,[4] that he was not the "first inventor" [5] of the telescope, some writers have insisted on attributing the invention to him.[6] But these opinions have never been in the center of historical scholarship and therefore need not concern us here.

The question as to the identity of the "certain Fleming" who had invented the telescope has exercised the minds of a number of writers on the subject. During the first half-century of the instrument's existence three names were put forward: Jacob Metius of Alkmaar,[7] Hans Lipperhey,[8] and Sacharias Janssen,[9]

[1] The persistent claim, based on Willem Boreel's reminiscences, that the compound microscope was invented by Sacharias Janssen and his father Hans, in Middelburg in 1590, gained prestige through the work of J. H. van Swinden (see G. Moll, "Geschiedkundig Onderzoek naar de eerste Uitvinders der Verrekijkers uit de Aantekeningen van wijlen den Hoogleeraar Van Swinden zamengesteld," *Nieuwe verhandelingen der eerste klasse van het Koninklijk Nederlandsch Instituut van wetenschappen, letterkunde en schoone kunsten* 3 [1831]: p. 171; *idem,* "On the first Invention of Telescopes," *Jour. Royal Inst.* 1 [1831]: pp. 319–332, 483–496). Pieter Harting, who had studied under Moll, became the most vociferous champion of this view (see e.g., *Het mikroskoop* [3 v., Utrecht, 1848–1850] 3: pp. 22–34, and especially the German editions of this work, e.g., *Das Microskop* [Braunschweig, 1859], pp. 585–595). De Waard discussed this claim in *De uitvinding der verrekijkers* (The Hague, 1906), pp. 293–304, and advocated the more reasonable view that the telescope was invented first and that the microscope evolved from it.

[2] The word *occhiale* was, and is, the Italian word for spectacles. The word *telescopium* was not coined until 1611. See Edward Rosen, *The Naming of the Telescope* (New York, Abelard-Schuman, 1947).

[3] I quote here from Stillman Drake's translation of *Sidereus nuncius,* in *Discoveries and Opinions of Galileo* (Garden City, Doubleday, 1957), pp. 28–29.

[4] Stillman Drake and C. D. O'Malley, *The Controversy on the Comets of 1618* (Philadelphia, University of Pennsylvania Press, 1960), pp. 211–213. See also Edward Rosen, "Did Galileo Claim he Invented the Telescope?" *Proc. Amer. Philos. Soc.* **98** (1954): pp. 304–312.

[5] Galileo referred to "the Hollander who was the first to invent the telescope" (Drake and O'Malley, *op. cit.,* p. 212). Pierre Borel called Sacharias Janssen *primus conspiciliorum inventor,* and Hans Lipperhey *secundus conspiciliorum inventor* (*De vero telescopii inventore* [The Hague, 1655–1656], captions of the portraits of Janssen and Lipperhey). In the English language too it was customary to refer to a "first inventor," see e.g., *Act 21, James I, c. 3, paragraph 6* (Statute of Monopolies, 1623–1624): "Lettres Patente . . . to the true and first Inventors of such Manufactures."

[6] E.g., Marcus Antonius de Dominis, *De radiis visus et lucis in vitris perspectivis et iride* (Venice, 1611), preface written by Giovanni Bartoli; Oliver Lodge, *Pioneers of Science* (London, 1893), pp. 95–96. Lodge claims that Lipperhey first invented the *astronomical* telescope and that, based on a rumor, Galileo then invented the "Dutch" or "Galilean" telescope.

[7] Jacob Adriaenszoon (d. 1628), usually referred to as Jacob Metius, was an instrument maker in Alkmaar. His father, Adriaen Anthonisz (*ca.* 1543–1620), several times a burgomaster of Alkmaar, was a military engineer and mathematician, whose approximation of π, 355/113, is still known as "the proportion of Metius." Jacob's brother, Adriaen (1571–1635) was professor of mathematics and astronomy at the University of Franeker. The origin of the family name "Metius" is not known. D. J. Struik recently wrote on this subject: "some derive it from Metz, others from the family name Schelven

both of Middelburg, and cases were built for the priority of each one.

The claim of Jacob Metius was systematically advocated by his brother Adriaen, professor of mathematics and astronomy at the Frisian university of Franeker. Adriaen Metius published a number of books on astronomy and geography, in both Latin and Dutch, and in these he claimed the invention for his younger brother Jacob.[10] Descartes followed the opinion of Adriaen Metius in his *Dioptrique* of 1637,[11] and thus brought the claim of Jacob Metius to the attention of a much wider reading audience.

But most writers thought the invention had originated not in Alkmaar, but rather in the city of Middelburg in the province of Zeeland. In his *Telescopium: sive ars perficiendi* of 1618 (but probably written as early as 1612),[12] a book entirely devoted to the so-called Dutch (or "Galilean") telescope, Girolamo Sirtori put forward the name of Ioannes Lippersein, a spectacle-maker of Middelburg. Sirtori also mentioned, however, that this spectacle-maker had apparently learned the art from a visitor who came to his shop and ordered a number of lenses.[13] Antonius Maria Schyrlaeus de Rheita simplified Sirtori's account and identified Ioannes Lippensum as the inventor of the telescope, in his *Oculus Enoch et Eliae* of 1645.[14]

The name of Sacharias Janssen was first mentioned in print in 1656 in a book devoted to the invention of the telescope, *De vero telescopii inventore,* written by Pierre Borel, one of the physicians of King Louis XIV of France.[15] Borel had enlisted the help of Willem Boreel, a native of Middelburg, and the ambassador of the Dutch Republic to the French Court.[16] Boreel had written to the City Council of his native city, requesting an investigation into the identity of the inventor, and the Council had sent him two sets of depositions, one in favor of Hans Lipperhey, and the other in favor of Sacharias Janssen.[17] On the basis of this information, as well as his own recollections—he had apparently been a playmate of Janssen during his early youth—Boreel decided that Sacharias Janssen was the inventor of the telescope.[18] Pierre Borel published these findings in his book, indicating his agreement with them.[19] *De vero telescopii inventore* put forward a plausible claim for the priority of Sacharias Janssen, and although it contained several serious inconsistencies,[20] it resulted in a general swing towards Janssen after 1656.

In 1682 Christiaan Huygens uncovered the first authentic document dating from a time very shortly after the invention of the telescope. It was a request by Jacob Metius to the States-General of the Netherlands for a patent on his invention of a new instrument for seeing faraway things as if they were nearby. But in this request, dated around 15 October, 1608, Metius mentioned that a citizen of Middelburg had already presented such an instrument to the States-General some time before him.[21] Huygens, however, never published this document. In his *Dioptrica* he did state his conclusion that Metius could not have been the inventor: it could have been either Lipperhey or Janssen.[22] But this work was not published until 1703, eight years after his death, and Huygens's opinion had little impact on scholarship dealing with this subject. Eighteenth-century authors often mentioned all three

(*schelf* = *rick* = Latin *meta*), it may also be related to *metiri* (to measure)." (*Dictionary of Scientific Biography* [New York, Charles Scribner's Sons, 1970–] **9**: p. 335.)

[8] Hans Lipperhey (d. 1619) was born in Wesel, and settled in Middelburg where he became a spectacle-maker. His name was variously rendered as Laprey, Lippershey, or Lipperhey.

[9] Sacharias Janssen, or Jansen, was probably born in The Hague around 1588. His father fled Antwerp in the 1580's for religious reasons and settled in Middelburg where he was a spectacle-maker. But he also traveled widely as a peddler. See de Waard, *op. cit.*, pp. 115–183.

[10] E.g., *Nieuwe geographische onderwysinghe* (Franeker, 1614) **15**; *Institutiones astronomicae et geographicae* (Franeker, 1614), pp. 3–4. See also p. 48, below.

[11] *Œuvres de Descartes publiées par Charles Adam & Paul Tannery* (Paris, 1897–1913) **6**: pp. 81–82. See also p. 53, below.

[12] See pp. 48–51, below.

[13] See pp. 48, 50, below.

[14] See pp. 53–54, below.

[15] See pp. 58–64, below. The full title of this work is *De vero telescopii inventore, cum brevi omnium conspiciliorum historia. Ubi de eorum confectione, ac usu, seu de effectibus agitur, novaque quaedam circa ea proponuntur. Accessit etiam centuria observationum microcospicarum* [*sic*]. Although the main title page gives the publication date as 1655 and the separate title page of the *centuria observationum* is dated 1656, the entire work was, in fact, issued in 1656. See A. Van Helden. "A Note about Christiaan Huygens's *De Saturni Luna Observatio Nova,*" *Janus* **62** (1975): pp. 13–15.

[16] The similarity of the names has often led to confusion. Willem Boreel (1591–1668), who latinized his name as *Borelius,* was a Dutch diplomat who served as ambassador in Paris from 1649 until his death. Pierre Borel (*ca.* 1620–1671), who usually latinized his name as *Borellus,* was a French physician, chemist, and antiquarian. In 1655 he was "medecin ordinaire" of King Louis XIV. Borel never became a member of the Académie Royale des Sciences, as is often stated. The physician who *did* become a member of that body in 1674 was Jacques Borelly (d. 1689), whose name was latinized as *Borellus* as well. None of these men must, of course, be confused with the Italian scientist Giovanni Alfonso Borelli (1608–1679), who latinized his name as *Borellus* or *Borellius.*

[17] See pp. 54–58, below.

[18] See pp. 63–64, below.

[19] See pp. 60–61, 63–64, below.

[20] These are discussed below, see pp. 23–24.

[21] See pp. 39, 40, below. This document was first published by Gerard Moll in 1831 ("Geschiedkundig Onderzoek," *op. cit.*, pp. 129–131), but the transcription used contained a number of important errors. Cornelis de Waard used the same faulty transcription (*Uitvinding,* pp. 23–24). The only reliable transcription is in *Œuvres complètes de Christiaan Huygens* (The Hague, 1888–1950) **13**: pp. 591–593.

[22] *Opuscula postuma* (Leiden, 1703), p. 163; *Œuvres complètes* **13**: pp. 436, 437.

candidates, sometimes giving the palm to Metius or Janssen, but never to Lipperhey.[23] This was the situation until the nineteenth century, when new light was shed on the subject.

In 1816 it was proposed at a meeting of the Zeeland Academy of Sciences to honor Sacharias Janssen by putting a commemorative stone in the gable of the house where he was thought to have lived.[24] This proposal led to an examination of Janssen's claim, and several contradictions were pointed out in the account given in *De vero telescopii inventore*.[25] A renewed search for original sources resulted in the discovery of a number of documents in the archives of the States-General in The Hague, dated between 2 October, 1608, and 13 February, 1609, dealing with the patent application of Hans Lipperhey.[26] Furthermore, independent evidence was found in the archives in Middelburg, showing that Lipperhey had indeed been a spectacle-maker in that city at the time of the patent application.[27] Nothing was found in these archives that substantiated the claim of Sacharias Janssen.

The men who wished to retain for Middelburg the honor of having produced the inventor of the telescope were now in a dilemma. The documents found in The Hague showed that Lipperhey had not been born in Middelburg: his native city was Wesel in Westphalia. It was firmly believed that Janssen had been born in Middelburg, and therefore his weakened claim was defended by the champions of Middelburg.[28] The eminent physicist Jan Hendrik van Swinden, however, was not so provincial, and on the basis of the new evidence he came out squarely in favor of Lipperhey in a lecture delivered in 1822.[29] Van Swinden died before he could present his researches in writing, but the astronomer Gerard Moll published a lengthy article based on van Swinden's notes, in one of the organs of the Royal Dutch Institute, in 1831,[30] in which the case for Lipperhey's priority was set out in a convincing manner. In the same year Moll published an English version of this article in the *Journal of the Royal Institution*.[31] On the basis of the evidence presented by Moll, the priority of Lipperhey was now generally accepted in the Netherlands as well as abroad. Lipperhey is, therefore, given the nod in most of the standard works on astronomy and optics published in the second half of the nineteenth century, although many of the writers do express some doubt and mention Sacharias Janssen also, especially in connection with the invention of the microscope.[32]

---

[23] In his "Recherche des Dates de l'Invention du Micrometre des Horloges à Pendule, & des Lunettes d'approche" (*Histoire de l'Academie Royale des Sciences. Année MDCCXVII. Avec les mémoires de mathématique & de phisique pour la même année* [Paris, 1719], Memoires, pp. 79–87), Philippe de la Hire did not identify the inventor and merely said that the instrument was invented by "le fils d'un ouvrier Hollandois" (pp. 84–85). This would tend to indicate Sacharias Janssen, however. In his first edition of *Histoire des mathématiques* (2 v., Paris, 1758), Jean-Étienne Montucla mentioned all three candidates, but concluded that Sacharias Janssen was the true inventor (2: p. 167). In the second edition of this work (4 v., Paris, 1799–1802) Montucla was more careful. After reviewing the known facts, he concluded: "Ce qui paroît en résulter sans difficulté, c'est que la ville de Middelbourg en Zélande est le lieu du berceau de cet admirable instrument, . . . dont l'inventeur n'est pas plus précisément connu; tel est le sort de presque toutes les découvertes les plus utiles à l'humanité" (2: p. 232). In the entry "Lunettes d'Approche" in Diderot's *Encyclopédie*, Metius is indicated as the inventor. Joseph-Jérôme le François de Lalande simply stated that the telescope was invented in Holland (*Astronomie* [Paris, 1771–1781] 2: p. 729). Jean-Sylvain Bailly mentioned all three candidates without deciding between them (*Histoire de l'astronomie moderne*, nouvelle edition [Paris, 1785] 2: pp. 84–85). Jean-Baptiste Joseph Delambre, however, mentioned only Metius in his *Histoire de l'astronomie moderne* (2 v., Paris, 1821) 1: pp. xix–xx: "La lunette avait été trouvée en Hollande, soit par un simple hasard, soit, comme il est plus probable, par les soins et la curiosité d'un amateur nommé Metius, dont le plaisir était de rassembler des lentilles de toute espèce et de les combiner ensemble pour en varier les effets." See also, Pieter De La Rue, *Geletterd Zeeland* (Middelburg, 1734), pp. 299–304.

[24] De Waard, *op. cit.*, p. 27.

[25] The contradictions will be discussed below. See pp. 23–24.

[26] See pp. 36–38, 42–44, below.

[27] De Waard, *op. cit.*, pp. 190–191.

[28] J. de Kanter and J. ab Utrecht Dresselhuis, *De provincie Zeeland* (Middelburg, 1824), Appendix X, pp. 79–98. P. Harting, "De twee Gewigtigste Nederlandsche Uitvindingen op Natuurkundig Gebied," *Album der Natuur* (1859), pp. 323–349, 355–368. Harting gives Lipperhey credit for the invention of the telescope but gives Janssen credit for the invention of the microscope. See also note 1, above. H. Japikse, *Het aandeel van Zacharias Janse in de uitvinding der verrekijkers* (Middelburg, 1890). Japikse concludes that Lipperhey invented the "Dutch" telescope, Janssen's father (Hans) invented the microscope, and Janssen himself invented the astronomical telescope!

[29] Moll, "Geschiedkundig Onderzoek," *op. cit.*, pp. 199–202.

[30] *Ibid.*, pp. 103–209.

[31] Moll, "On the First Invention of Telescopes," *op cit.*

[32] E.g., Robert Grant, *History of Physical Astronomy* (London, 1852), p. 519: "No mention of Jansen could be found in any of the State Records, and as his right to the original invention of the telescope is set aside by the authentic documents in favour of Lipperhey, there only remains to be claimed for him the invention of the microscope, upon the evidence set forth in Borel's book. We may conclude, then, that Lipperhey was the person who originally executed telescopes, and also that he was the first who made them known to the world; and, therefore, upon these grounds he possesses a just claim to the honour associated with the invention." In *Die erfindung des fernrohrs und ihre folgen fuer die astronomie* (Zurich, 1870), Rudolph Wolff mentions only Lipperhey but feels that the real credit for the invention of the telescope ought to be given to Kepler (pp. 7–8). In his *Geschichte der astronomie* (Munich, 1877), however, he mentions both Lipperhey and Janssen but does not decide between the two. He does give Janssen the credit for inventing the microscope (pp. 358–359). H. Servus discusses the issue at length in *Die geschichte des fernrohrs bis auf die neuste zeit* (Berlin, 1886), pp. 5–42. After discussing Moll's Dutch paper of 1831 (pp. 38–42), he concludes: "Als Erfinder des Fernrohrs ist also allein Lippershey und 1608 als Jarezahl der Erfindung anzusehen." A Berry, in *A Short History of Astronomy* (London, 1898), states: "The effective discovery of the telescope was made in

In 1906 the most comprehensive work on the invention of the telescope ever undertaken was published by Cornelis de Waard. In *De uitvinding der verrekijkers* de Waard presented a thorough review of the issue as well as a large number of new documents which shed fresh light on the controversy. Where others had been unable to find any information concerning the life of Sacharias Janssen, de Waard was able to write an almost complete biography of this elusive man, based on documentary evidence.[33] Janssen, it turned out, was also not a native of Middelburg; most likely he was born in The Hague.[34] Moreover, this man whom the city of Middelburg had proudly claimed as one of its illustrious (if somewhat obscure) sons, turned out to be a very unsavory character. At various times he was in trouble with the law for non-payment of debts (among them sizable drinking bills), for assaults on fellow citizens, and for counterfeiting Spanish coins. For this last offense he was eventually threatened with the death penalty.[35] While many would consider this inability to stay within the bounds of the law reason to discount any legitimate role which Janssen might have played in the invention or dissemination of the telescope, de Waard found in it precisely the characteristics which suited Janssen ideally for the role he was to play.[36]

The central piece of information uncovered by de Waard was an entry in the unpublished journal of Isaac Beeckman, the rector of the Latin school in Dordrecht, and a friend of Descartes.[37] Beeckman learned how to grind lenses for telescopes in an effort to obtain better instruments. In the early 1630's he took lessons from a spectacle-maker in Middelburg named Johannes Sachariassen, the son of Sacharias Janssen. Beeckman recorded in his journal that during one of these lessons Johannes Sachariassen told him that the first telescope in the Netherlands had been made in 1604 by his father, after the model of an instrument in the possession of an Italian. This instrument bore the date 1590.[38] According to de Waard, it would not be in the nature of Sacharias Janssen, an entrepreneur with scant respect for the law, who knew from experience how easily the instrument could be copied, to seek a patent. Instead, he would have tried to offer it for sale to important persons in order to make money out of it quickly.[39]

De Waard concluded that the telescope was not invented in the Netherlands after all, but that its origins should be sought in Italy around 1590.[40] On the basis of some rather dubious evidence he also concluded that Raffael Gualterotti of Florence had been in possession of a telescope before the turn of the seventeenth century.[41]

The researches of de Waard were summarized in *Ciel et Terre* in 1907,[42] and his book was discussed at length by Antonio Favaro.[43] The new evidence thus found its way into the literature.[44] Lipperhey's claim was now seriously weakened, even though no clear candidate was left in his absence. But in the English-speaking world de Waard's researches had little immediate impact. Here Lipperhey remained the prime candidate, although the name of Janssen was still frequently mentioned on the basis of *De vero telescopii*

Holland in 1608 by Hans *Lippersheim* (?–1619), a spectacle-maker of Middelburg, and almost simultaneously by two other Dutchmen, but whether independently or not it is impossible to say" (p. 149).

33 De Waard, *Uitvinding,* pp. 115–183.

34 *Ibid.,* pp. 117, 323, 330.

35 Janssen was prosecuted for counterfeiting twice, once in Middelburg in 1613 and once in nearby Arnemuiden in 1618–1619. In the first instance the municipal court in Middelburg merely fined him; in the second case the provincial authorities, the States of Zeeland, took an active interest in the case and instructed the complacent authorities of Arnemuiden to prosecute Janssen to the full extent of the law. This case was dropped when Janssen disappeared (*ibid.,* pp. 120–138). In my article "The Historical Problem of the Invention of the Telescope," (*History of Science* **13** (1975): pp. 251–263), p. 257, I implied that Stillman Drake was in error in calling Janssen a "convicted counterfeiter" (*Galileo studies: Personality, Tradition, Revolution* [Ann Arbor, University of Michigan Press, 1970], p. 155). From the above, it is clear that when Janssen was assessed a fine in 1613 he was, in fact, formally found guilty, and therefore it is I who was in error, not Professor Drake.

36 De Waard, *Uitvinding,* pp. 162–163. De Waard does mention that perhaps Janssen offered telescopes to Prince Maurice and Archduke Albert and was asked to keep the device a secret (pp. 161–162).

37 De Waard later published Beeckman's journal in its entirety: *Journal tenu par Isaac Beeckman de 1604 à 1634 publié avec une introduction et des notes par C. de Waard* (4 v., The Hague, Martinus Nijhoff, 1939–1953).

38 *Ibid.,* 3: p. 376. See also p. 53, below.

39 De Waard, *De uitvinding der verrekijkers,* p. 162.

40 *Ibid.,* p. 157.

41 *Ibid.,* pp. 97–104.

42 De Waard, "L'Invention du Télescope," *Ciel et Terre* **28** (1907): pp. 81–88, 117–124.

43 A. Favaro, "La Invenzione del Telescopio secondo gli ultimi Studi," *Atti del Reale Istituto Veneto di scienze, lettere ed arti* **66**, 2 (1907): pp. 1–54.

44 The most important reference is in A. Danjon and A. Couder, *Lunettes et télescopes* (Paris, Éditions de la Revue d'Optique théorique et Instrumentale, 1935), pp. 583–604. Through this classic work, the reference to de Waard's researches entered the literature on this subject in France, as well as some other countries. In Italy the article by Favaro (see note 43, above), was important, and Vasco Ronchi has referred the reader to de Waard, beginning with his *Galileo e il cannocchiale* (Udine, Casa Editrice Idea, 1942), pp. 132–134, and has mentioned de Waard's conclusions in all his subsequent books on optics. In Germany, de Waard's work has had little impact. Ernst Zinner does not mention it in either *Entstehung und ausbreitung der coppernicanischen lehre* (Erlangen, 1943), or *Deutsche und niederlaendische astronomische instrumente des 11.—18. jahrhunderts* (Munich, C. H. Beck, 1958). In his *Fernrohre und ihre meister* (Berlin, D.D.R., V.E.B. Verlag Technik, 1957), pp. 21–22, Rolf Riekher has recourse only to Moll's 1831 article and Borel's 1656 publication.

inventore.[45] Not until the last twenty years has the existence of *De uitvinding der verrekijkers* been generally known in the English and American schools of history of science, thanks mainly to the translation of such works as Antonie Pannekoek's *De groei van ons wereldbeeld*[46] and Vasco Ronchi's various books on the history of light and optics.[47]

Since the publication of *De uitvinding der verrekijkers* in 1906, little additional evidence has been uncovered. Dutch scholars have traced the peripatetic Janssen to Amsterdam during the last years of his life,[48] but no further light has been shed on the events before 1608. It now appears that we may never know the identity of the inventor(s) of the telescope,[49] and that all we can hope for is a better understanding of the state of the art in the lens-grinding profession around the turn of the seventeenth century and its relationship to the more esoteric writings on optics of men like Giovanbaptista Della Porta and his contemporaries.

II. THE BACKGROUND

Although it is now generally agreed that the telescope could not have been invented much before 1600, it seems appropriate to review here some of the earlier events—not so much in order to discount some of the extravagant claims that have been made (e.g., that Roger Bacon knew about the telescope),[1] as to sketch some of the background against which the events around 1600 unfolded.

Using a tube without lenses to aid vision was a well-known practice in Antiquity. In his *Generation of Animals,* Aristotle wrote:

> The man who shades his eye with his hand or looks through a tube will not distinguish any more or any less the differences in colours, but he will see further; at any rate, people in pits and wells sometimes see the stars . . . [780 b 15, ff.].
>
> . . . distant objects would be seen best of all if there were a sort of continuous tube extending straight from the sight to that which is seen, for then the movement which proceeds from the visible objects would not get dissipated; failing that, the further the tube extends, the greater is bound to be the accuracy with which distant objects are seen [781 a 5 ff.].[2]

Early writers on the telescope, such as Johann Baptist Cysat (1586–1657), were sometimes led by references to, and indeed illustrations of, such tubes to ascribe the invention and use of the telescope to the Ancients.[3] There is no doubt, however, that tubes such as the ones incorporated in the spheres of Gerbert of Aurillac (*ca.* 945–1003), who became Pope Sylvester II, were not telescopes but rather tubes without lenses, sometimes called "polar sighting tubes."[4]

---

[45] E.g., Louis Bell, *The Telescope* (New York, McGraw-Hill, 1922), pp. 2–7; Peter Doig, *A Concise History of Astronomy* (London, Chapman & Hall, 1950), pp. 66–67; Henry C. King, *The History of the Telescope* (London, Charles Griffin, 1955), pp. 30–32. King discusses the claims of Lipperhey, Metius, and Janssen, and concludes that there is still doubt as to the identity of the inventor (p. 30). He does refer to Danjon and Couder (*loc. cit.*) with regard to Janssen's claim, and therefore my statement in "The Telescope in the Seventeenth Century" (*Isis,* **65**: p. 39, n. 2), that King seems unaware of de Waard's conclusions, is incorrect. King does not, however, mention the entry in Beeckman's journal and the fact that Janssen may have been in possession of an instrument as early as 1604. In the foreword of this book, the entry is mentioned by Sir Harold Spencer Jones (vii). Reginald S. Clay and Thomas H. Court, in *The History of the Microscope* (London, Charles Griffin, 1932), pp. 7–9, do not mention de Waard's work and refer the reader back to Moll's article in *Jour. Royal Inst.* (*loc. cit.*).

[46] Antonie Pannekoek, *De groei van ons wereldbeeld* (Amsterdam, Antwerp, Wereld-Bibliotheek, 1951), pp. 184, 205; *idem, A History of Astronomy,* no translator indicated (New York, Interscience Publishers; London, George Allen and Unwin, 1961), pp. 227, 253.

[47] Vasco Ronchi, *Histoire de la lumière,* tr. J. Taton (Paris, Armand Colin, 1956), p. 83; *idem, The Nature of Light,* tr. V. Barocas (London, Heinemann, 1970), p. 95.

[48] J. C. van Breen, "Topographische Geschiedenis van de Dam te Amsterdam," *Zevende jaarboek der vereeniging Amstelodamum* (Amsterdam, 1909), pp. 183, 188. H. F. Wijnman, "Sacharias Jansen te Amsterdam," *Amstelodamum* **20** (1933): pp. 125–126; *idem,* "Nogmaals Sacharias Jansen," *ibid.* **21** (1934): pp. 82–83; C. W. van der Sterr, "Hans en Zacharias Janssen," *ibid.* **21** (1934): pp. 70–71. See also J. H. Kruizinga, "De Strijd om een Verrekijker," *Historia. Maandschrift voor geschiedenis en kunstgeschiedenis* **13** (1948): pp. 140–144, and Bruno Ernst, "Een Standbeeld voor Zacharias Janssen," *Zeeuws tijdschrift* **25,** 2 (1975): pp. 29–36.

[49] The municipal archives of Middelburg, in which so many documents bearing on this issue were found between 1816 and 1906, were housed in the city hall. When the city hall was bombed by the *Luftwaffe* in 1940 the archives were entirely destroyed. Today Middelburg's city hall stands again, the product of a remarkable reconstruction effort. The oldest document in its present archive dates from 1940.

[1] See, e.g., William Molyneux, *Dioptrica nova, a treatise of dioptricks* (London, 1962), pp. 256–257; S. Jebb, ed., *Fratris Rogeri Bacon, ordinis minorum, opus maius* (London, 1733), 9th and 10th unnumbered pages of the *praefatio;* [John] D[rinkwater], "Curious Extracts from Old English Books, with Remarks which prove, that the Telescope &c. were known in England much earlier than in any other Country," *Philos. Mag.* **18** (1804): pp. 245–256; **19** (1804): pp. 66–79, especially pp. 77–78; R. T. Gunther, *Early Science in Oxford* (14 v., Oxford, 1921–1945) 2: pp. 288–289. After quoting the passage from Bacon's *Epistola de secretis operibus* (see p. 28, below), Gunther writes: "It is inconceivable that man should have arrived at such a thought without some practical experience, or without information of the practical experience of others."

[2] Aristotle, *Generation of Animals with an English Translation by A. L. Peck* (London, Heinemann; Cambridge, Harvard University Press [Loeb Classical Library], 1943), pp. 503, 505.

[3] Johann Baptist Cysat, *Mathemata astronomica de loco, motu, magnitudine, et causis cometae qui sub finem anni 1618 et initium anni 1619 in coelo fulsit* (Ingolstadt, 1619), p. 76.

[4] Gerbert's letter on this subject can be found in *The Letters of Gerbert with his Papal Privileges as Sylvester II. Translated with an introduction by Harriet Pratt Lattin* (New York,

Before the telescope could be invented, lenses had to exist. Precious stones which were ground into lens-like shapes may, from time to time, have been used in Antiquity as magnifiers or even as visual aids to help a person with defective sight see better. But it was not until the medieval period in Europe that discs of non-precious stones were ground into convex shapes for the specific purpose of helping people with defective vision. This invention of spectacle lenses has been a subject of historical investigation since the seventeenth century, and Edward Rosen has recently undertaken an exhaustive review of the accumulated scholarship on this subject. His investigations show that, if we cannot be sure of the identity of the inventor(s), it is nevertheless clear that the invention occurred in the last few decades of the thirteenth century.[5]

These first convex eyeglasses were designed to help old people in reading. The condition known today as *presbyopia* is due to the progressive loss of the accommodating power of the human eye with age. When fully relaxed, the normal human eye sees distant objects sharply. For seeing objects nearer than about 20 meters away, the refractive power of the optical system of the eye must be increased, and this is done by contracting the cilia muscles around the lens, which increases the curvature of the lens. At age ten the normal eye can accommodate itself to see objects as near as three inches from the eye in sharp focus. This "near point" recedes with age, and in the middle forties it is about ten inches removed from the eye. At this point a person begins to experience difficulties with close work, such as reading. When the "near point" has reached arm's length, say twenty inches (so that a person cannot see anything nearer than twenty inches from the eye in sharp focus), reading becomes virtually impossible without optical aids, and progressively stronger lenses are required to allow older people to read. Early convex eye-glasses, then, were a boon to aging scholars, extending their useful reading and writing life substantially.

Another visual defect, usually associated with reading, is the condition known as *myopia*. In this case the eyeball is elongated (along the eye's optical axis), and the refractive power of the eye is too great to allow a focusing of light rays from distant objects on the retina. The result is that a myope can see nearby objects sharply but cannot see remote objects with clarity. The ophthalmologist Julius Hirschberg (1843–1925) found in a sample of 7,573 people that 2 per cent of the peasants were myopic while 16 per cent of the clerks and 32 per cent of the university students were myopic.[6] Obviously, then, myopia is a condition closely associated with the proportion of his time a person spends in reading and writing, and its incidence could conceivably be used as a rough indicator of literacy.

The number of books published annually increased by several orders of magnitude as a result of the invention of printing with movable type. Accordingly, the incidence of myopia increased rapidly after about 1450, and the correction of this defect by means of concave spectacle lenses came about some time around this date. Until recently, the earliest known references to concave spectacle lenses were from the middle of the sixteenth century,[7] although many scholars speculated that the mention of concavely ground beryl, in Nicholas of Cusa's *De beryllo* of 1458 meant that Cusa was familiar with such lenses.[8] Recently, however, Vincent Ilardi has published five previously unknown documents, found in diplomatic correspondences of the fifteenth century, which prove conclusively that concave spectacle lenses were for sale in Florence by 1451 and could be bought in quantity by 1462.[9] This evidence, moreover, points to Florence as the place of origin of these eyeglasses which were ordered from places such as Ferrara and Milan.[10] We may therefore date the use of concave lenses for aiding the vision of myopes to around 1450. It was not until some time later that they became common items in the shops of spectacle-makers in other Italian cities, and it appears that they spread slowly to the rest of Europe from Italy, where, according to Girolamo Mercuriale (1530–1606), the

Columbia University Press, 1961), pp. 36–39. See also S. Guenther, "Das Glaeserlose Sehrohr im Altertum und Mittelalter," *Bibliotheca mathematica* (published by Gustaf Enestroem, Stockholm), N.S., **8** (1894): pp. 15–23; Robert Eisler, "The Polar Sighting-tube," *Archives internationales d'histoire des sciences* 2 (1949): pp. 312–332; Henri Michel, "Les Tubes Optiques avant le Telescope," *Ciel et Terre* **70** (1954): pp. 175–184. Eisler and Michel show a number of contemporary illustrations of these instruments.

[5] Edward Rosen, "The Invention of Eyeglasses," *Jour. History of Medicine and Allied Sciences* **11** (1956): pp. 13–46, 183–218. See also Rosen's articles "Did Roger Bacon Invent Eyeglasses?" *Archives internationales d'histoire des sciences* 7 (1954): pp. 3–15; and "Carlo Dati on the Invention of Eyeglasses," *Isis* **44** (1953): pp. 4–10, reprinted in Otto Mayr, ed., *Philosophers and Machines* (New York, Science History Publications, 1976), pp. 82–88.

[6] Hirschberg, *The Treatment of Shortsight,* tr. G. Lindsay Johnson (London, Rebman, 1912), p. 4.

[7] Vasco Ronchi, "A Fascinating Outline of the History of Science. Two Thousand Years of Conflict Between 'Reason' and 'Sense,'" *Atti della Fondazione Giorgio Ronchi* **30** (1975): p. 530.

[8] In the 1488 Strasburg edition of Cusa's *Opera,* the beginning of the second paragraph of *De beryllo* reads: "Berillus lapis est lucidus albus & transparens: cui datur forma concava pariter & convexa & per ipsum videns attingit prius invisibile intellectualibus oculis." See also Rosen's comment on this passage "The Invention of Eyeglasses," p. 206, note 290. See also H. J. M. Weve, "Brillen fuer die Ferne," chapter XI in R. Greeff *et al., Katalog einer bilderausstellung zur geschichte der brille* (Amsterdam, A. E. D'Oliveira, 1929), pp. 44–48.

[9] Vincent Ilardi, "Eyeglasses and Concave Lenses in Fifteenth-Century Florence and Milan: New Documents," *Renaissance Quarterly* **29** (1976): pp. 341–360.

[10] *Ibid.*

incidence of myopia was higher than elsewhere.[11] They probably did not become common in northern Europe until the middle of the sixteenth century.

Although we can thus point to the presence of both concave and convex lenses in Europe by, say, 1525, and to their common presence in the inventory of spectacle sellers a generation or so later, we must keep in mind that the mere presence of both types of lenses was by no means a sufficient condition for the invention of the telescope. In order to get a telescopic effect, one has to put together two lenses of suitable focal lengths. This raises the difficult question: what was the range of strengths of lenses made and sold by spectacle-makers? In trying to answer this question, we must at all times keep in mind that the quality of glass was very low and that lens-grinding techniques were too primitive to give a glass disc a uniform spherical curvature. The resulting lenses were, therefore, very poor by our standards.

The correction of presbyopia was, until the middle of the nineteenth century, a straightforward procedure. When a person could no longer read without the aid of spectacles, he or she went to a spectacle-maker's shop and bought a pair of reading glasses. As early as the seventeenth century there was a rule of thumb relating the strength of the reading glasses required to a person's age.[12] The first and weakest glasses had focal lengths of 12 to 20 inches (i.e., strengths of 2 to 3 diopters).[13] These were by far the most common reading glasses: in Italy they were called *occhiali di vista commune*.[14] Stronger convex glasses were available for older people and people who had had cataract operations. Since there was no call for weaker lenses (which were, at any rate, more difficult to grind, as is abundantly clear from early seventeenth-century efforts to make objective lenses for longer telescopes), it is very unlikely that one would encounter convex lenses with focal lengths of more than about 20 inches in the shops of spectacle-makers around 1600.

Much less is known about the correction of myopia by means of concave lenses around the turn of the seventeenth century. Myopia was, in most places, not as common as presbyopia, and since there was no easy rule of thumb to guide spectacle-makers, it seems unlikely that a myope could walk into a spectacle-maker's shop and be fitted properly with "ready-made" spectacles. It must have required a more laborious process, often resulting in the artisan making several pairs to try to satisfy the customer, and we may surmise that not all common spectacle-makers could (or would) do this. Grinding strong concave lenses was a laborious and difficult task, and it is not unreasonable to assume that spectacle-makers learned only slowly how to make stronger concave lenses. Presumably the strongest concave lenses to be found in a spectacle-maker's shop around 1600 had focal lengths of 8 to 12 inches.

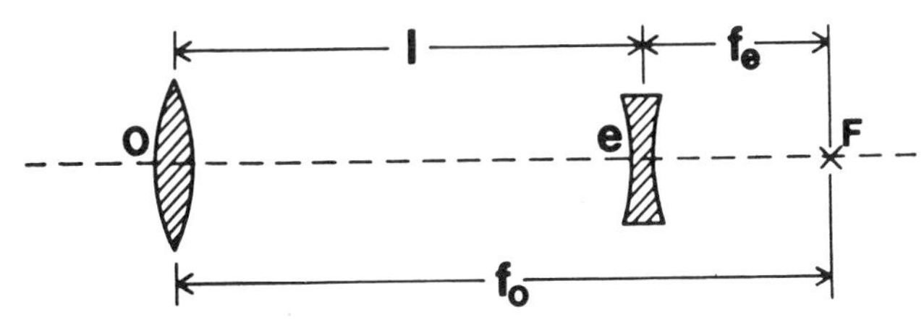

FIG. 1. Optical system of a "Galilean" telescope. O is the objective lens, e is the eyepiece. F is the focal point of the objective as well as the focal point of the eyepiece. f_o is the focal length of the objective; f_e is the focal length of the eyepiece; l is the length of the telescope.

If, now, we were to combine the weakest convex lenses with the strongest concave lenses found in shops around the turn of the seventeenth century, we could make a telescope which magnified 2 or 2½ times. The time was rapidly approaching when such combinations of lenses would have an obviously useful magnification.

The above speculation is supported by the information available on very early telescopes. In April, 1609, Pierre l'Estoile (1546–1611) examined telescopes of about a foot in length at a spectacle-maker's shop in Paris.[15] In August of that same year Giovanbaptista Della Porta described a 14-inch telescope in a letter to Federigo Cesi.[16] Let us say, then, that the very first telescopes made were 12 to 14 inches long, and that these instruments were made up of lenses found in a spectacle-maker's shop. Galileo remarked in his *Sidereus nuncius* that the first telescope he made (put together, obviously, from lenses he bought in a shop) magnified three times.[17] Combining this information, we can calculate the focal lengths of the lenses used in these early telescopes.

Figure 1 presented here shows the optical system of a "Galilean" telescope. O is the convex objective and e the concave eyepiece. In a "Galilean" telescope the lenses are arranged as shown, with the eyepiece positioned so that its back focal point coincides with the focal point of the objective, F. f_o is the focal length of the objective, f_e the focal length of the eyepiece, and l the length of the telescope—the eye is placed immedi-

[11] J. Stilling, *Untersuchungen ueber die entstehung der kurzsichtigkeit* (Wiesbaden, 1887), pp. 182–183, 190.

[12] Benito Daça de Valdés, *Uso de los antojos* (Seville, 1623). I have used an anonymous French translation made in 1627, first published in Giuseppe Albertotti, ed., "Manoscritto Francese del Secolo Decimosettimo riguardante l'Uso degli Occhiali," *Memorie della Regia Accademia di scienze, lettere ed arti in Modena,* ser. 2, **9** (1893): pp. 27–37. Carlo Antonio Manzini, *l'Occhiale all'occhio* (Bologna, 1660), p. 98.

[13] Daça de Valdés, *op. cit.,* pp. 36–37; Manzini, *op. cit.,* p. 98.

[14] Manzini, *loc. cit.*

[15] See p. 44, below.

[16] See pp. 44–45, below. A Naples *palmo* was 10.378 in. and an *oncia* 0.862 in., according to W. B. Parsons (*Engineers and Engineering in the Renaissance* [Cambridge, M.I.T. Press, 1968], p. 629). This would make this instrument about 14 inches long.

[17] *Sidereus nuncius,* in Stillman Drake, *Discoveries and Opinions of Galileo* (Garden City, Doubleday, 1957), p. 29.

ately behind the eyepiece. From the diagram it follows that

$$l = f_o - f_e \quad (1)$$

We know, furthermore, that the magnification m of a simple telescope is given by the ratio of focal lengths,

$$m = f_o/f_e \quad (2)$$

Solving these equations for $m = 3$, and $l = 12$ and $l = 14$ inches, we obtain for a 12-inch telescope $f_o = 18$ inches and $f_e = 6$ inches, and for a 14-inch telescope $f_o = 21$ inches and $f_e = 7$ inches. Accordingly the weakest convex lenses found in a spectacle-maker's shop in 1609 had focal lengths of 18 to 21 inches, and the strongest concave lenses had focal lengths of 6 to 7 inches. These figures agree well with our speculations, and we may conclude that around the turn of the seventeenth century it had just become possible to construct a weak but usable Galilean telescope from lenses available in the shops of spectacle-makers.

The unresolved problem of the actual genesis of the instrument has given occasion to many speculations concerning the meaning of many of the obscure utterings on lenses and mirrors by sixteenth-century authors. It is necessary, at this point, to re-examine this literature in the light of the foregoing discussion.

When Roger Bacon wrote "from an incredible distance we might read the smallest letter and number grains of dust and sand," [18] he was expressing an attitude towards nature which may seem admirable to the modern reader, but which was regarded with suspicion by the church authorities of his time because it smacked of magic. We may be fairly certain that Bacon was not familiar with spectacles—they were, at any rate, invented some time after the *Opus majus* was completed—[19] and the claims that he actually knew the workings of a telescope are, of course, unfounded.

From their invention in the late thirteenth century until well into the sixteenth century, lenses were mentioned only rarely in philosophical literature. They remained craft items which did not attract the attention of writers on optics, because these scholars worked within the classical Greek tradition of dealing with refraction through entire spheres. Vasco Ronchi has dramatically called this lack of attention to lenses a "conspiracy of silence." [20] He explains it by arguing that it was generally thought that since lenses distort reality, they could not help the investigator arrive at the truth of Nature.[21] David Lindberg and Nicholas Steneck have recently questioned this explanation, arguing that Ronchi based his view on insufficient evidence.[22] Regardless of the merits of the arguments on either side of this particular issue, the fact remains that scholars showed little or no interest in lenses until the sixteenth century.

In contrast to the essentially modern point of view, which would see Friar Bacon as a misunderstood prophet, and the lack of interest in practical and mundane things on the part of the medieval schoolmen as an historical aberration which needs explaining, recent scholarship has concentrated on explaining the increasing interest in practical things on the part of sixteenth-century thinkers. The bridge between the theoretical and the practical—between contemplative knowledge and knowledge applied to more practical ends—is to be found, according to these scholars, in the various aspects of magic, which experienced a revival in the Renaissance. Through magic they see the essential manipulative element entering the preoccupations of European natural philosophers during this period.[23] This view is in good agreement with the evidence in the case of lenses and mirrors. We find the first references to the powers of lenses and mirrors in the writings of such Renaissance *magi* as John Dee (1527–1608) and Giovanbaptista Della Porta (*ca.* 1535–1615). Only after the invention of the telescope does the study of lenses and lens systems become an important part of formal optics.

There are, then, increasingly frequent references to the possibilities of lenses and mirrors in the second half of the sixteenth century. The earliest ones are obscure and fanciful, and do not represent advances over the speculations of Friar Bacon, to whom these writers frequently refer.[24] Quite naturally, these early writers

[18] See p. 28, below.

[19] Rosen, "The Invention of Eyeglasses," *op. cit.*

[20] Vasco Ronchi, *The Nature of Light,* tr. V. Barocas (London, Heinemann, 1970), p. 71. For a list of early references to and pictorial representations of spectacles, see George Sarton, *Introduction to the History of Science* (3 v., Baltimore, Williams & Wilkins, 1927–1948) 2: p. 1026.

[21] Ronchi, *op. cit.,* p. 73.

[22] David C. Lindberg and Nicholas H. Steneck, "The Sense of Vision and the Origins of Modern Science," in: Allen G. Debus, ed., *Science, Medicine and Society in the Renaissance. Essays to Honor Walter Pagel* (2 v., London, Heinemann, 1972) 1: pp. 29–45. See also Lindberg's review of *The Nature of Light,* in: *Isis* 62 (1971): pp. 522–524. For Ronchi's reply to Lindberg, see "How the History of Science Should be Approached," *Atti della Fondazione Giorgio Ronchi* 29 (1974): pp. 47–72, and "A Fascinating History," *loc. cit.* (note 7 above).

[23] In the introduction to his monumental *History of Magic and Experimental Science* (8 v., New York, Columbia University Press, 1923–1958) 1: p. 2, Lynn Thorndike wrote: "My idea is that magic and experimental science have been connected in their development; that magicians were perhaps the first to experiment." Recent studies of Renaissance magic have all started from Frances A. Yates's *Giordano Bruno and the Hermetic Tradition* (London, Routledge & Kegan Paul, 1964). An interesting attempt to do justice to the role of "the magical tradition" in the Scientific Revolution can be found in Hugh Kearney, *Science and Change, 1500–1700* (New York, McGraw-Hill, 1971).

[24] E.g., Robert Recorde, *The Pathway to Knowledge* (London, 1551), preface; see p. 29, below. John Dee possessed

concentrated on the references to miraculous devices in the literature, such as Archimedes' burning mirrors "that dyd bourne their ennemies shyppes a great way from the town," and Bacon's "glasse that he made in Oxforde, in which men myght see thynges that were doon in other places." [25] They started out by trying to find out "howe a great numbre of them were wroughte, that they may be practiced in this tyme also" [26] and therefore these early references to what could be accomplished by means of "art perspective" are not to be taken as indications of the existence of such wonderful devices.

On the other hand, optical devices such as the *camera obscura* were being used by this time, and one begins to see a different attitude towards lenses in, e.g., the statements of such sober sages as Girolamo Fracastoro (1478–1553) [27] and other physicians, such as Jacques Houllier (Hollerius) (d. 1562), who were making the correction of various forms of "weak vision" by means of spectacles a legitimate concern of doctors.[28] Already in John Dee's "very fruitfull praeface" to Henry Billingsley's English translation of Euclid's *Elements* (1570) we find an explicit admission of the limitations of the contemporary abilities which, nevertheless, does not undermine Dee's faith in the potentialities of optical devices: the military tactician "may wonderfully helpe him selfe by perspective Glasses. In which, (I trust) our posterity will prove more skillful and expert, and to greater purposes, then in these dayes, can (almost) be credited to be possible." [29] The lesson is clear: the fanciful and the practical went hand in hand, and if there was a separation, it was only one of time —today's fancy would be tomorrow's practice.

It is difficult to think that for all the fanciful statements about "numbering peeces of money with the very coyne and superscription thereof," [30] Leonard Digges (d. *ca.* 1571) and his son Thomas (*ca.* 1543–1575) did not actually do some experiments with combinations of lenses and mirrors.[31] The writings of William Bourne (fl. 1565–1588) make this even clearer. His discussions are not limited to fanciful devices; they also tell the reader how to proceed practically.[32] His *Treatise on the properties and qualities of glasses for optical purposes,* written in the middle 1580's, is a very useful review of the state of the art and is therefore included in its entirety in the document section.[33] Bourne gives us an important piece of information in his introduction to the chapter on lenses:

> And nowe furdermore, as I have shewed before, the forme, and facyon of glasses, that dothe reflect a beame from the glasse, commonly called Lookinge Glasses; So in lyke manner I will shewe you the makinge of Glasses called perspective glasses, that do help sighte, by meanes of the beame, that pearceth commonly thorowe the glasse. And first for the makinge of the smallest sorte of them, commonly called spectacle glasses.[34]

The word "glasses" could be used, then, to denote both lenses and mirrors, and if one wanted to be more specific one would use the terms "looking glass" for mirror and "perspective glass" for lens. Throughout Bourne's treatise, the term "glass" by itself refers to a mirror. Since there is not a single reference in the tract to concave lenses, we may be certain that the term "perspective glass" referred exclusively to convex lenses. Harriot's "perspective glasse whereby was shewed manie strange sightes" [35] was, therefore, nothing more than a magnifying glass. When telescopes were actually introduced into Harriot's circle, in 1609, they were referred to as "perspective cylinders" or "perspective trunks." [36]

Bourne describes convex lenses in a way which leaves no doubt as to his familiarity with them: when one is close to them they magnify slightly, and the magnification increases as one moves farther away from them, until a point is reached at which one can "discerne nothing thorowe the glasse: But like a myst, or water; And at that distance ys the burninge beame." Beyond this point everything is seen "cleene turned, and reversed another way." [37] His description of convex and concave mirrors is equally complete. But when it comes to the combination of concave mirrors and convex lenses (which, he claims, will enable one to see things far away), he pleads poverty and lack of time and refers the reader to Mr. Dee and Mr. Thomas Digges. Whereas before he had spoken in definite

numerous manuscript copies of Bacon's writings. The 1618 Hamburg edition of *De secretis operibus* was taken from Dee's manuscript version.

[25] Recorde, *The pathway to knowledge, loc. cit.*

[26] *Ibid.*

[27] See p. 28, below.

[28] Jacques Houllier, *De morbis internis* (Paris, 1571), Liber I, 76v–81v. See also, Singer, *op. cit.*, p. 400.

[29] See p. 29, below.

[30] See p. 30, below.

[31] Thomas did write that his father engaged in some sort of experiments, or, as he called them, "continual painfull practices" (*ibid.*).

[32] William Bourne, *Inventions or devices. Very necessary for all generalles and captaines, or leaders of men, as well by sea as by land* (London, 1578), pp. 96–97. The 110th device, a gadget with which to see faraway things (see p. 30, below) is a good example of this mixture of the fanciful and the practical. The entire book is in this vein.

[33] See pp. 30–34, below.

[34] See p. 32, below.

[35] See p. 34, below. In *Astronomical Thought in Renaissance England* (Baltimore, Johns Hopkins Press, 1937), Francis R. Johnson states that all these references are to "some form of early telescope" (p. 178).

[36] A useful review of the terminology can be found in John North, "Thomas Harriot and the first Telescopic Observations of Sunspots," in: John W. Shirley, ed., *Thomas Harriot, Renaissance Scientist* (Oxford, Clarendon Press, 1974), pp. 143–144. But I disagree with North's conjecture that the "perspective glasse" in Harriot's *Briefe and true report* of 1588 was probably a mirror (p. 143).

[37] See p. 33, below.

terms, leaving no doubt as to the extent of his personal experience with these matters, now he takes on a much more tentative tone:

> But notwithstanding upon the smalle proofe and experyence those that bee but unto small purpose, of the skylles and knowlledge of these causes, yet I am assured that the glasse that ys grounde, beynge of very clear stuffe, and of good largenes, and placed so, that the beame doth come thorowe, and so reseaved into a very large concave lookinge glasse, That yt will shewe the thinge of a marvellous largenes, in manner uncredable to bee beleeved of the common people.[38]

Or, as Roger Bacon had put it three hundred years earlier, *"ut animus mortalis ignorans veritatem non posset sustinere."* [39]

There is no doubt that one can achieve a magnifying effect with a combination of an appropriate convex lens and concave mirror. But it is very doubtful that Bourne had actually experimented with such combinations. Had he done so, he would have known that the beam from the convex "perspective glass" should be received in a very *small* concave "looking glass" (i.e., one with a short focal length), not in a very *large* one. Combining a very large concave mirror with a very large convex lens would result in no magnification. What Bourne was thinking, one may surmise, was that both convex lenses and concave mirrors enlarge things, and if you combine the two the magnifying effect would be additive: if one "glass" magnifies, two will magnify more. But this is not the principle of the telescope, and to ascribe to Bourne a knowledge of the telescope on the basis of such utterings is wrong.

Most probably, Bourne was referring to a passage in *Pantometria* published in 1571 by Thomas Digges, but written mostly by his father, Leonard Digges. Digges's reference to "glasses concave and convex of circulare and parabolicall fourmes, using for multiplication of beames sometime the ayde of glasses transparent, whiche by fraction should unite or dissipate the images or figures presented by the reflection of other" [40] has been interpreted as a description of a reflecting telescope.[41] But, tantalizing as it is, it does not warrant such a conclusion. The entire passage is vague and obscure, and the "marveylouse conclusions" which may be obtained with such a device are clearly fanciful. There is no mention of the fact that parabolic mirrors were impossible to make at this time, that foil-backed mirrors of glass would lead to hopeless results, and there is no hint as to what sort of mirror, concave or convex, is to receive the light, or what sort of mirror or lens is to serve in the capacity of what we would call the eyepiece.

What is interesting about this passage is the reference to a transparent glass which dissipates the images—a concave lens. To my knowledge, in the entire body of English literature on "perspective" in the second half of the sixteenth century this is the only reference to concave lenses. It was not followed up by the other writers of the English school, perhaps because concave lenses were still rare in England at this time, but also because concave lenses make things appear smaller. Such lenses would find no place in speculations about the possibility of compounding the effect of magnifying glasses or mirrors.

The conclusions to be drawn from the writings of Robert Recorde, Leonard and Thomas Digges, John Dee, and William Bourne are that, on the basis of the available writings on formal optics and the fanciful utterings of Friar Bacon, these "mathematical practitioners" speculated about the powers of lenses and mirrors, and combinations thereof, becoming gradually less fanciful and more practical in their writings. They never considered a combination of two or more lenses, always referring to combinations which included at least one mirror, and, moreover, neglected the possibility of using concave lenses. In working with mirrors, they were limited to foil-backed glass mirrors, which would add two refractions to one reflection, and we can be fairly certain that, in view of the difficulty of shaping glass (not to speak of the poor quality of the glass itself) the resulting images could not have been very useful, regardless of whether a lens or yet another mirror was used to compound the magnifying effect.

In general, their notion was that the miraculous effects mentioned by Roger Bacon were to be achieved by compounding the effect of one magnifier by adding other magnifiers. Concave lenses, which make things appear smaller, were therefore ignored. If Digges or Bourne had actually been in possession of a combination of mirrors and/or lenses which constituted what we would call a reflecting telescope, surely the instrument would quickly have been adopted for military use. Can one seriously think that William Bourne, with his "greate charge of children" and his meager "hability of

[38] See p. 34, below.

[39] See p. 28, below.

[40] See p. 29, below.

[41] R. T. Gunther, *Early Science in Oxford, op. cit.* 2: pp. 289–291. Besides quoting the passage (p. 290) from the 1571 edition of *Pantometria* (p. 29, below), Gunther also quotes the following passage (pp. 289–290): "By concave and convex mirrors of circular (spherical) and parabolic forms, or by paires of them placed at due angles, and using the aid of transparent glasses which may break, or unite, the images produced by the reflection of the mirrors, there may be represented a whole region; also any part of it may be augmented, so that a small object may be discerned as plainly as if it were close to the observer, though it may be as far distant as the eye can descrie." His source for this quotation is (p. 290, note 1): *"Pantometria,* 1571. The exact wording varies in different editions." I have been unable to find this particular quotation in the editions of *Pantometria* which I have consulted. But the wording (e.g., mirrors for glasses) is suspiciously modern, and it appears that Dr. Gunther has quoted the same passage twice, once in the original form found in the 1571 edition of *Pantometria* (p. 290), and once in modern English (pp. 289–290).

. . . purse," would not have tried to capitalize on such a very useful device? We may be sure that the "perspective glasse" of Thomas Harriot, and the *specillis* with which William Gilbert looked at the Milky Way,[42] were nothing more than simple magnifying glasses, as Bourne's explanation of the terms would indicate.

If the English practitioners ignored concave lenses and combinations consisting of lenses only, things were different in Italy. Girolamo Fracastoro had already mentioned a combination of two spectacle lenses, one placed on top of the other, in 1538, although there is no evidence to indicate that one of these was a concave lens.[43] His countryman Giovanbaptista Della Porta, the best known of the Renaissance *magi,* wrote in the second edition of his immensely popular *Magia naturalis* (1589):

> With a Concave you shall see small things afar off, very clearly; with a Convex, things neerer to be greater, but more obscurely: if you know how to fit them both together, you shall see both things afar off, and things neer hand, both greater and clearly.[44]

This passage has been interpreted by many scholars, beginning with Kepler,[45] as an unmistakable (although obscure) reference to the telescope. As it stands, the passage is indeed seductive. But the picture changes if we consider it in its proper context. Book XVII of *Magia naturalis* deals with mirrors and lenses. The first nine chapters are devoted entirely to mirrors, describing plain mirrors, burning mirrors, and trick mirrors. Chapter 10 is the first chapter which deals with lenses. After a discussion of burning lenses (in which Porta mentions that burning lenses kindle fire more quickly than burning mirrors) he turns to the magnifying effects of these convex lenses. One can read a letter a great way off by means of these glasses, and if the lens is inclined to the direction of vision, the letters will appear even larger—one can see them at twenty paces. By multiplying this effect through the use of two or more of these lenses, one could see the same letters at a hundred paces.[46] Then comes the famous passage, which begins:

> Concave Lenticulars will make one see most clearly things that are afar off; but Convexes, things neer hand; so you may use them as your sight requires.[47]

This is nothing but a simple statement about spectacle lenses: concave lenses are to be used by people who cannot see things that are far away; convex lenses by people who cannot see things that are near. The passage which supposedly describes the telescope follows immediately. Let us now read the entire passage:

> Concave Lenticulars will make one see most clearly things that are afar off; but Convexes, things neer hand; so you may use them as your sight requires. With a Concave you shall see small things afar off, very clearly; with a Convex, things neerer to be greater, but more obscurely: if you know how to fit them both together, you shall see both things afar off, and things neer hand, both greater and clearly.

Porta is speaking here strictly about the *correction* of defective vision, not about the *extension* of normal vision. If there could be any doubt about this, it is dispelled by the very next sentence:

> I have much helped some of my friends, who saw things afar off, weakly; and what was neer, confusedly, that they might see all things clearly.[48]

Moreover, if Porta were referring to an instrument to be used by people with normal vision to see things that are beyond their normal range of vision, would one not expect him to make extravagant claims for the powers of such a device? But this is precisely what he does in the very next chapter, entitled "Of Spectacles whereby one may see very far, beyond imagination." In order to put Porta's writings in their proper perspective, it is useful to quote this passage in full:

> I will not omit a thing admirable and exceeding useful; how bleare-ey'd people may see very far, and beyond that one would believe. I spake of *Plotomies* [*sic*] Glass, or rather spectacle, whereby for six hundred miles he saw the enemies ships coming; and I shall attempt to shew how that might be done, that we may know our friends some miles off, and read the smallest letters at a great distance, which can hardly be seen. A thing needful for mans use, and grounded upon the Opticks. And this may be done very easily; but the matter is not so to be published too easily; yet perspective will make it clear. Let the strongest sight be in the Centre of the Glass, where it shall be made, and all the Suns beams are most powerfully disperst, and unite not, but in the Centre of the foresaid Glass: in the middle of it, where diameters cross one the other, there is the concourse of them all. Thus is a Concave pillar-Glass made with sides equidistant: but let it be fitted by those Sections to the side with one oblique Angle: but obtuse Angled Triangles, or right Angled Triangles must be cut here and there with cross lines, drawn from the Centre, and so will the spectacle be made that is profitable for that use I speak of.[49]

---

[42] William Gilbert, *De mundo sublunari philosophia nova* (Amsterdam, 1651), p. 250. Gilbert died in 1603. This work was published from his manuscript by William Boswell.

[43] See p. 28, below.

[44] The translation is from *Natural Magick by John Baptista Porta, a Neapolitane: in twenty books* (London, 1658), p. 368. See p. 35, below.

[45] Kepler, *Dissertatio cum nuncio sidereo* (1610), see Edward Rosen, *Kepler's Conversation with Galileo's Sidereal Messenger* (New York, Johnson Reprint, 1965), pp. 15–18. See also Rosen's notes, pp. 79–83.

[46] See p. 35, below.

[47] *Ibid.*

[48] *Ibid.*

[49] *Magia naturalis,* p. 270 (the misspelling of Ptolemy is in the English translation only); *Natural Magick,* p. 369. See also Edward Rosen's discussion of Porta's claim to the invention of the telescope in *The Naming of the Telescope* (New York, Henry Schuman, 1947), pp. 6–23.

The reader is invited to fathom Porta's meaning here. It is certain, however, that he was not thinking of a combination of lenses. Porta, like Digges and Bourne, undoubtedly had first-hand experience of lenses and mirrors and was interested in miraculous devices such as those ascribed to Roger Bacon. In 1589 he did not have knowledge of a telescope. Had he had such knowledge, he would not have kept it entirely secret (although he would probably not have published it), but would have tried to exploit it much as Lipperhey and Galileo exploited the first actual telescopes.

But although there is no evidence to suppose that the telescope was known before about 1600, lenses were coming into practical use more and more during the second half of the sixteenth century. There is no doubt that Fracastoro had seen the Moon and the stars through a magnifying glass,[50] that Harriot and Gilbert used "perspective glasses" for various purposes, and that Kepler saw Halley's Comet in 1607 *per perspicilla,* which were probably ordinary spectacles.[51] Military commanders were counseled by Leonard Digges to use "perspective glasses," and there is good reason to believe that such glasses were becoming more and more common. Moreover, during the second half of the sixteenth century glass-making techniques pioneered in Italy spread through Europe, often carried abroad by the Italian glass-makers themselves.[52] Better glass, so necessary for optical devices, was becoming available, and lens-grinding techniques were being steadily improved (although they were still at a very low level).

And perhaps the utterings of writers on "perspective" supplied an impetus to grind convex lenses with longer focal lengths. In his treatise, Bourne describes a perspective glass which is of interest in this connection. If one wanted to make an excellent lens, one should,

> prepare very cleare, and white Glasse that may be rounde, and beare a foote in diameter; as fyne and white Vennys Glasse. And the larger, the better: and allso yt must bee of a good thickness, and then yt must bee grounde uppon a toole fitt for the purpose. Beynge sett fyrst uppon a syman [i.e., cement] block, and first, grynde on the one syde, and then on ye other syde, untill that the sydes bee very thynn, and the middle thicke. And for that yf the glasse bee very thicke, then yt will hynder the sighte. Therefore yt must bee grounde untill that the myddle thereof bee not above a quarter of an ynche in thickness: and the sydes or edges very thynne, and so polysshed or cleared. And so sette in a frame meete for the purpose for use: so that yt may not be broken.[53]

Assume that the curvature of the surfaces of such a lens is equal on both sides and that the thickness of the lens decreases smoothly from one-quarter inch in the middle to zero at the edges. Then the minimum radius of curvature of the surfaces would be twelve feet. Such a lens could not be ground a hundred years later in any figure useful for optical purposes! Note also that the transparency of even the best Venice glass was apparently so poor that, if the lens was not ground to a minimum thickness, the passage of light through it would be severely impeded. We may assume, therefore, that, if a lens of these dimensions had indeed been made by Bourne, it would have been entirely useless.

Nevertheless, the passage from Bourne's treatise indicates that perhaps there was a growing demand for lenses with longer focal lengths, made of glass of the highest quality. How far practice lagged behind Bourne's desires is amply illustrated by the comments on Lipperhey's first telescope in 1608. We may safely assume that this instrument had an objective lens with a focal length of two feet or less and a diameter of at most two inches. Yet it was so poor that Lipperhey was asked by the States-General to make the lenses of his next telescopes out of rock crystal.[54]

By about 1600 we can point to a steadily improving glass-making and lens-grinding industry, and to the presence of both convex and concave lenses in the shops of spectacle-makers. There was also a well-established belief in the possibility that some combination of lenses and/or mirrors would produce miraculous magnifying effects, and there was a growing demand for convex lenses with longer focal lengths. The turn of the seventeenth century, then, marks the threshold at which the fanciful ideas of the Renaissance *magi* could be translated into practical devices. The result was not long in coming.

III. BETWEEN PORTA AND LIPPERHEY, 1589–1608

If we can agree that none of the writers discussed in the previous chapter had an actual Galilean telescope (at least of noticeable and useful magnification), the question as to when and how the telescope was invented, some time between 1589 and 1608, still remains. Although it is impossible to document the actual invention, there are, nevertheless, some pertinent observations to be made regarding the genesis of the instrument. These observations have to be based on the instrument itself.

From time to time writers have wondered why the so-called Galilean telescope (consisting of a convex objective lens and a concave eye lens) was invented before the so-called Keplerian version (consisting of two convex lenses). The Keplerian version—later called the astronomical telescope—yields an inverted image, and it could be that such a combination of lenses was, in fact, found very early but discarded

[50] See p. 28, below.
[51] See p. 35, below.
[52] R. J. Charleton and L. M. Angus-Butterworth, "Glass," in: Charles Singer, *et al., A History of Technology* (5 v., Oxford, Clarendon Press, 1954–1958) **3**: pp. 216–219.
[53] See p. 33, below.
[54] See p. 36, below.

because of this inverted image. There is, however, a very plausible explanation of why the Galilean configuration was found first. This explanation is to be found in the actual optics of the two possible combinations.

How might a person experimenting or playing with lenses have discovered the magnifying effects of these two possible combinations? In trying to reconstruct this process, we must begin with what someone familiar with lenses, e.g., Porta, knew:[1]

(1) Convex lenses enlarge things but make them indistinct.
(2) Concave lenses show things sharply defined but smaller.
(3) The stronger the glass, the greater the magnifying or diminishing effect.

Let us suppose now that we have at our disposal the entire range of lenses—concave and convex—that might be found in a spectacle-maker's shop around 1610. If we pick out a strong magnifier, say a lens with a focal length of 6 inches, and hold it right up against our eye, we see a distant object, say a weather cock, right-side-up at about its normal size, but slightly blurred. Moving the glass away from our eye, we see the weather cock progressively larger and more blurred, until the glass is about 6 inches from our eye. At that point, all definition is gone; in the words of William Bourne, we can "discerne nothing thorowe the glasse: But like a myst, or water: And at that distance ys the burninge beame."[2]

When, now, we move the glass a little farther away from the eye, the weather cock can again be made out. It appears magnified, blurred, and "reversed and turned the contrary way."[3] As we continue to move the lens farther from the eye, the inverted image of the weather cock becomes progressively sharper and smaller, until at about 12 inches from the eye the glass shows the inverted image in sharp focus and of the same size as the weather cock seen with the naked eye. Moving the lens beyond this point, we see the weather cock inverted, in sharp focus, and progressively smaller until it becomes too small to discern. We know now, therefore, that a single convex lens will not give us an image that is *both* enlarged and in sharp focus.

In spite of this fuzziness of the image, which William Bourne ascribed to the poor quality of lenses,[4] our natural inclination would be to try to compound the magnifying effect of one convex lens by adding another convex lens. Obviously we would begin by adding another *strong* magnifier, say another 6-inch glass. If we put the two lenses on top of each other and hold them up to our eye, we again see the distant weather cock erect, at about its normal size, but much more blurred than through a single lens. When we keep one against the eye and begin moving the other one towards the weather cock, the image becomes larger and more blurred until at about 6 inches separation between the lenses nothing is seen through them. When the separation is increased slightly, an enlarged, blurred, inverted image appears, and as the separation increases this image becomes smaller and sharper. When the separation of the lenses is about twelve inches the weather cock again appears in its normal size, sharply defined, but inverted. Further increase in the separation between the two lenses leads to a rapid loss of definition while the image becomes smaller. The desired effect cannot be obtained in this manner.

Should we decide to try a combination of a strong convex and a weak convex—and such a combination is anything but obvious—we should have to decide which of them to put next to the eye. If we choose to hold the weaker one against the eye and move the stronger one towards the weather cock, we would obtain roughly the same result as we did with the single strong magnifier. If, however, we hold the strong lens—the 6-inch one—against the eye and begin moving the weak one—say an 18-inch lens—away from the eye towards the distant weather cock, the result would be different. The erect, fuzzy, magnified image of the weather cock would increase in size and fuzziness until the separation between the lenses was about 18 inches. At this point, as before, all definition would be lost. A further increase in the separation would bring into view a highly distorted, inverted and magnified image which, at a separation of 24 inches, rather suddenly loses its distortion (since the magnification is somewhat less). At this point we see the weather cock sharply defined, inverted, and enlarged about three times. A further increase in the separation between the lenses leads to a very rapid increase in fuzziness of the image.

Here is, then, the effect we are looking for—a compounding of the effect of a single magnifying glass. But notice that the separation of 24 inches puts the weaker lens at extreme arm's length for a tall person. At that distance it is very difficult to keep the lenses aligned properly. In view of the fact that in this case the desired effect appears suddenly and is equally suddenly lost, this difficulty in handling the lens at arm's length is an important factor. The effect is somewhat easier to find with an (objective) lens which

[1] See pp. 34–35, below.
[2] See p. 33, below.
[3] *Ibid.*
[4] Nowhere in his tract (pp. 30–34, below) does Bourne say that convex lenses make things appear indistinct or fuzzy. He appears to think that the fuzziness is due to the poor quality of the available glass, for he repeatedly tells the reader that the glass must be "of a very cleare stuffe" such as "fyne and white Vennys Glasse" in order to achieve the effects which he is supposedly describing.

is a bit stronger—say 12 inches in focal length—but in that case the magnifying effect of the combination is considerably less. Nor could we, in that case, use a stronger convex eye lens, for we would not be able to find such lenses with a focal distance of less than about 6 inches in a spectacle-maker's shop at this time.

Besides the fact, therefore, that the Keplerian configuration gives an inverted image, its effect is also difficult to find by means of experimenting, not to mention playing with lenses. There are no common-sense indications as to what particular combination of convex lenses to choose, or which one should be the object lens and which one the eye lens, and the desired effect can be found only within a very narrow range of distances between the two lenses. It is by no means an obvious combination.

Consider now concave lenses. Taking one concave lens and putting it before the eye, we see a sharply defined, erect field, slightly diminished in size. As we move the lens away from the eye, the image of the weather cock remains erect and sharp but becomes progressively smaller, no matter how far we move the lens away from the eye. Looking through two concave lenses, we find that the effect is simply additive, no matter what the separation between the lenses. There is, thus, no way we can achieve magnification using two concave lenses.

Why, however, should we ever consider using a concave lens in any combination in an effort to achieve the desired effect? After all, concave lenses make things appear smaller! The answer lies in the function of concave spectacle lenses. They are used by myopes, i.e., people who cannot see things that are far away: "Concave Lenticulars will make one see most clearly things that are afar off." [5]

If then we begin to experiment with combinations of concave and convex lenses, there are again several possibilities. Looking at the distant weather cock through a strong convex lens placed on top of a weak concave lens, we see the image larger and erect, but blurred. Keeping the convex lens against our eye and moving the concave lens away from the eye, we see the weather cock get progressively smaller and fuzzier until we can no longer distinguish anything. And no matter how far we increase the separation, the weather cock will not appear again.

Holding the weaker concave glass near our eye and moving the stronger convex glass towards the weather cock, we see the blurred, erect image grow larger and fuzzier until at the distance of the "burning beam" of the convex all definition is gone. Beyond that point a slightly enlarged, inverted weather cock becomes visible again (much the same as with a single convex lens removed from the eye by slightly more than its focal distance) until it is sharply defined at about twice the focal distance of the convex lens. The image is slightly smaller at this point than the weather cock seen with the naked eye because of the diminishing effect of the concave lens near the eye. Further increase in the separation between the lenses will just make this image smaller. A combination of a strong convex and a weak concave glass will thus not give us the effect we are looking for.

Now we try a combination in which the concave lens is stronger than the convex one. Let us pick a 6-inch concave glass and an 18-inch convex glass. Holding both lenses up to our eye together, we see a sharply defined and erect weather cock which is slightly diminished in size. Keeping the convex lens against our eye and moving the concave lens away from it, we see the weather cock become smaller and smaller without interruption, until it is too small to see. But it remains sharp throughout.

Finally, we hold the stronger concave lens against the eye and begin moving the weaker convex glass away. The weather cock which was seen erect, sharply defined, and slightly smaller than normal when both lenses were together against the eye remains sharply defined and erect and becomes progressively larger until the lenses are about 12 inches apart. At that point the weather cock is seen three times as large as its normal size. A slight increase in the separation will cause a total loss of definition, and a still further increase will slowly bring an inverted view of the weather cock into focus. This image will diminish as the distance is increased further. This combination *will* give us the effect we are looking for.

The point of this somewhat tedious discussion is that whereas it is rather difficult to discover the telescopic effect of a Keplerian configuration of lenses, it is very easy to discover it in the Galilean configuration. Starting with the appropriate combination of lenses placed on top of each other, when one moves them apart, the image *which is sharply defined and erect at all times* expands until it disappears. Moreover, one would naturally tend to hold the concave lens against the eye because through it one will at all times see a sharply defined field in which the convex lens itself is clearly visible. If we put ourselves in the shoes of Porta, it is impossible to argue that men such as he were not aware of this effect! But how, then, can we explain the historical facts?

In *De uitvinding der verrekijkers* Cornelis de Waard pursued the trail pointed out by the statement concerning Sacharias Janssen to Italy.[6] If the word of Janssen's son was to be trusted, we should look for the invention of the telescope in Italy. He quoted several writers who, he thought, had knowledge of the telescope but concluded that it was not possible to pin-

[5] See p. 35, below.

[6] See p. 53, below.

point the time and place of the actual invention.[7] The telescope was there, he wrote, before anyone knew it.[8] But de Waard only dimly perceived how accurate a statement this, in fact, was!

Although we are obviously aware of the enormous difference between, say, the 200-inch Hale telescope and Galileo's early efforts preserved in Florence, we nevertheless call them both "telescopes" because conceptually they are the same to us. Taking this conceptual identification back into the prehistory of the telescope, however, is an historical error. The identification only becomes pertinent when the aspiration of the *magi* to produce Bacon's wonderful glass and the mundane achievements of optical craftsmen merged in the first decade of the seventeenth century. Looking back from that vantage point we must recognize the crucial separation between these two strands.

If we try to assess the role of men such as Porta in the genesis of the telescope, the important question is: what were they looking for? And the simple answer to that question is that they were looking for a truly miraculous device (with an effect like that of the 200-inch Hale telescope), not for something mundane that might magnify slightly. In this matter, as in others, they drew their inspiration from Roger Bacon: "Thus from an incredible distance we might read the smallest letters and number grains of dust and sand." [9] Porta speculated on how one might make a glass such as King Ptolemy had made, a glass "whereby for six hundred miles he saw his enemies ships coming." [10] But, as shown above, this speculation is not closely related in his book to his discussion of the mundane combination of a concave and a convex lens.[11]

In view of the above discussion of combinations of lenses, and also in view of Porta's own writing in Book XVII, Chapter 10 of the 1589 edition of his *Magia naturalis,* we can no longer maintain that Porta had not put a concave glass together with a convex glass and that he was unaware of the effect of this combination. What we *can* argue is that the lenses he worked with—lenses he got from a spectacle-maker—gave him combinations which magnified very little, say 1.5 or 2. He used such combinations (probably at less than their greatest magnification) to help friends with certain visual defects, but he never saw in them a way to begin approaching the effects that Friar Bacon had written about.

What, then, would we expect Porta's reactions to be when, twenty years later, he saw one of the new instruments about which he had been hearing so much for some time? In August, 1609, shortly after examining one of the much touted new spyglasses for the first time, he wrote to Federigo Cesi: "About the secret of the spectacles [*occhiali*], I have seen it, and it is a hoax [*coglionaria*], and it is taken from the ninth book of my *De refractione.*" [12] The book *De refractione* had been published by Porta in 1593, and as a matter of fact it does not contain any mention of a combination of a concave and a convex lens. He probably made an error and meant to write "the seventeenth book of my *Magia naturalis.*" But the important point about Porta's reaction here is that he does not so much argue that someone had stolen the invention from him as attack the invention itself. He states that the invention is a hoax, that is, *it is not an invention at all!* This is precisely the reaction we would expect from one who had long been familiar with such combinations but had always thought them unimportant because he considered the effect trivial when compared with the effects ascribed to the "glasses" of Roger Bacon and King Ptolemy.

And Porta was not the only one who had such a trivial combination of lenses before 1608. In April, 1610, shortly after the publication of *Sidereus nuncius,* Raffael Gualterotti (1548–1639) wrote to Galileo:

> It is now twelve years ago that I made an instrument, but not for the purpose of seeing great distances and measuring the stars, but rather for the benefit of a cavalry soldier in joust and warfare. And I offered it to the Most Serene Grand Duke Ferdinand, and at the same time to the Most Illustrious and Excellent Lord Duke of Bracciano, Don Verginio Orsino. But as it seemed to me a feeble thing, I neglected it.[13]

In 1598, therefore, Gualterotti had made a combination like the one described by Porta in 1589 to help a soldier see in joust and warfare. Obviously a gadget with appreciable magnification would be counterproductive at close quarters such as in joust. This device must, therefore, have had a very low magnification, so low, in fact, that neither Ferdinand de'Medici nor Verginio Orsino showed any interest in it or recognized its potential. Even Gualterotti himself thought it a "feeble thing" and forgot about it.

Technically, according to *our* conception of the instrument, a device consisting of a concave and a convex lens separated by some distance and with the concave lens stronger than the convex lens (as it must have been in order to yield sharply defined images), could be considered a "telescope." But neither Porta nor Gualterotti considered these spectacles in any way related to the mythical "glasses" of King Ptolemy and Roger Bacon. In that sense, therefore, "telescopes" existed before anyone, *including the men who made them,* were aware of them.

Between Porta's 1589 publication and Lipperhey's patent application in 1608 several things happened. Concave lenses became much more common outside

[7] *De uitvinding der verrekijkers,* pp. 92–104.
[8] *Ibid.*, p. 91.
[9] See p. 28, below.
[10] See p. 15, above.
[11] See pp. 15–16, above.
[12] See p. 44, below.
[13] See pp. 45–46, below.

Italy, while the spectacle-makers were also learning to grind these lenses with shorter focal lengths in order to correct more severe cases of myopia. There may also have been a slight increase in the focal lengths of convex lenses. The effect of putting together a convex lens and a stronger concave lens and drawing them apart as one holds the concave glass against the eye is so easily noticed, as shown above, that it could not have been any particular secret. One might go so far as to say that it must have been a rather common piece of knowledge among people working with lenses.

At some point lenses were available in the shops of spectacle-makers which when combined in this fashion had a magnification of more than about two. Now it was only a matter of time before someone would recognize that such a combination could be very useful and begin trying to improve it. At precisely *this* point the fanciful notions about miraculous devices for seeing faraway things were grafted onto a particular combination of lenses which had been known for some time. In this confluence of the tradition of Friar Bacon and the anonymous craft tradition the modern concept of the telescope was born. Its birth certificate (now lost) is Lipperhey's patent application of October, 1608, its first cry was Galileo's *Sidereus nuncius* of March, 1610, and its baptism was the Lyncean feast on the Malvasia estate on 14 April, 1611.[14]

[14] Edward Rosen, *The Naming of the Telescope* (New York, Abelard-Schuman, 1947).

IV. LIPPERHEY, METIUS, AND JANSSEN

If the *prehistory* of the telescope occurred mainly in England and Italy, at least so far as the available sources go, its *history* began in the Netherlands—specifically, in the two western provinces, Holland and Zeeland. The Dutch Republic was a federation of seven provinces governed by the States-General in The Hague. Each province had its own governing body, e.g., the States of Zeeland, made up of delegates mostly from the cities. When these provincial states were not in session, routine business was handled by committees known as the *Gecommitteerde Raden,* or Committees of Councillors. The seat of the government of the province of Zeeland was the city of Middelburg on the island of Walcheren. It was the most important commercial and manufacturing center in Zeeland which flourished after the fall of Antwerp to the Spanish in 1585 and the subsequent closing of the Schelde, Antwerp's link with the sea, by the Dutch navy.

On 25 September, 1608 (by the Gregorian Calendar), the Committee of Councillors of the Province of Zeeland wrote a letter to one of Zeeland's delegates to the States-General in The Hague. This letter instructed the delegate to recommend its bearer (who is not identified) to Prince Maurice, Count of Nassau, stadholder of five of the seven provinces, and commander in chief of the Dutch armed forces.[1] The bearer, they wrote, claimed to have "a certain device by means of which all things at a very large distance can be seen as if they were nearby, by looking through glasses which he claims to be a new invention." [2] Exactly a week later an entry was made in the minute book of the States-General describing the patent application of Hans Lipperhey, a native of Wesel, and now a citizen of Middelburg, where he was a spectacle-maker.[3]

The ensuing documents describe the dealings between Lipperhey and the States-General. A committee was appointed to investigate the instrument and to negotiate with the inventor. He was asked to construct binocular telescopes so that one would not have to look through one eye and close the other, to use rock crystal (i.e., quartz) instead of glass in their construction, and to deliver six instruments within a year. Lipperhey's initial demand for 1,000 guilders for each instrument was considered excessive.[4] On 5 October, Lipperhey was paid 300 guilders as an advance and promised another 600 guilders upon the satisfactory completion of the work. The decision on his patent application was postponed until that time.[5] On 15 December the committee reported that it had examined a binocular telescope made by Lipperhey.[6] At this meeting of the

[1] Maurice, Count of Nassau (1567–1625), was the second son of William the Silent. He became stadholder of the provinces of Holland and Zeeland in 1585, the year after his father's death, and at that time he was also appointed commander in chief of the armed forces of these two provinces, a position which was extended to include the armed forces of all seven provinces in 1589. In 1590 he became stadholder of the provinces Utrecht, Gelderland, and Overyssel as well. Not until 1618, when his older half-brother Philip William died, did the title Prince of Orange accrue to him. Because of his military successes against the Spanish forces Maurice attained a very high military reputation.

[2] Pp. 35–36, below.

[3] P. 36, below.

[4] *Ibid.*

[5] Pp. 37–38, below.

[6] P. 42, below. From the ensuing negotiations it appears that the committee interpreted six instruments to mean three binocular instruments. Note also that Lipperhey was thus the first to make a binocular telescope. Antonius Maria Schyrlaeus de Rheita was the first to claim this invention in print in his *Oculus Enoch et Eliae* of 1645 (part 1, p. 337). Giovanbattista Clemente de Nelli claimed this invention for Galileo in *Saggio di storia letteraria fiorentina del secolo xvii* (Lucca, 1759), pp. 70–72, and in *Vita e commercio letterario di Galileo Galilei* (2 v., Lausanne, 1793) **1**: pp. 280–283; **2**: pp. 680–683. Although Antonio Favaro was unaware that Lipperhey had made a binocular telescope as early as 1608, he did correct Nelli: the *celatone* made by Galileo in 1617 was not a binocular telescope but a helmetlike devise to which a single telescope was firmly attached in an attempt to make possible the observation of Jupiter's satellites (which required high magnification) on board ship. Favaro also claimed that Ottavio Pisani may have made a binocular telescope as early as 1613 ("Sulla Invenzione dei Cannocchiali Binoculari," *Atti della*

States-General the patent application was rejected, but Lipperhey was asked to make two more binocular instruments and given another 300 guilders. The remaining 300 guilders of the 900 originally promised were to be paid on the delivery of these two instruments.[7] On 13 February, 1609, it was recorded that Lipperhey had delivered these last two instruments and the final payment of 300 guilders was ordered.[8] Lipperhey received the money on the same day.[9] No further entries concerning Lipperhey are to be found in the minutes or the account book of the States-General. Thus, he did not get his patent, but he did earn the princely sum of 900 guilders in less than five months for making three binocular instruments.

Why was Lipperhey denied a patent on such an obviously useful invention? The answer to this question can be found in contemporary documents. In the minutes of the Committee of Councillors of Zeeland for 14 October, 1608, there is an entry which states that these gentlemen interviewed another person who claimed to know the art of making spyglasses.[10] That same day they wrote a letter to their representative in The Hague, alerting him to the fact that "there is here a young man who says that he also knows the art, and who has demonstrated the same with a similar instrument." [11] Their advice was that the instrument could not be kept a secret:

> we believe that there are others as well, and that the art cannot remain secret at any rate, because after it is known that the art exists, attempts will be made to duplicate it, especially after the shape of the tube has been seen, and from it has been surmised to some extent how to go about finding the art with the use of lenses.[12]

Although we cannot be certain of the identity of this young man, it is reasonable to assume, as Cornelis de Waard did, that it was Sacharias Janssen who was about twenty years old at that time and was also a spectacle-maker in Middelburg.[13]

A letter written some time around the fifteenth of October throws yet a third hat into the ring. This is the application for a patent for the same instrument by Jacob Adriaenszoon of Alkmaar (a city in the north of the province of Holland), usually known as Jacob Metius. Metius wrote that he had been investigating the powers of lenses for two years and that (presumably at some point during those two years) he had invented an instrument for seeing faraway things. He had concentrated all his efforts on improving the instrument from then onward and had finally brought it to a reasonable state of perfection. His instrument was as good, at any rate, as that which had recently been shown to the States-General by a spectacle-maker from Middelburg. Thus, Metius had heard about Lipperhey's patent application and had hastened to put in his claim: he had found the instrument first and was about to be robbed of the rightful fruits of his labor.[14]

Within three weeks of the original letter from the Committee of Councillors of Zeeland, therefore, we find the telescope in the possession of three men. But there is good reason to suppose that the story is even more complicated. In his *Mundus jovialis* of 1614 Simon Marius (1570–1624) relates how he first became aware of the existence of the new instrument. His patron Johann Philip Fuchs von Bimbach was at the autumn fair in Frankfurt a.M. in 1608 when he heard,

> that there was then present in Frankfurt at the fair a Dutchman, who had invented an instrument by means of which the most distant objects might be seen as though quite near. . . . Our nobleman [Fuchs] had a long discussion with the Dutch first inventor, and felt doubts as to the reality of the new invention. At last the Dutchman produced the instrument, which he had brought with him, and one glass of which was cracked, and told him to make a trial of the truth of his statements. So he took the instrument into his hands, and saw that objects on which it was pointed were magnified several times. Satisfied of the reality of the instrument, he asked the man for what sum he would produce one like it. The Dutchman demanded a large price, and when he understood that he could not get what he first asked, they parted without coming to terms.[15]

Although Marius has been accused of lying about his observations of Jupiter's satellites, there is no reason to question his integrity here too. By the autumn of *1609* spyglasses were commonly for sale in the major cities of Europe [16] and it would be hard to believe that only one person was trying to sell one at Europe's foremost fair in that year. Marius could, therefore, not have been mistaken in the year: it must have been 1608. Now, officially the autumnal fair began on 15 August and ended on 8 September every year.[17] But it appears that, in fact, it began early in September and lasted perhaps a month.[18] At the latest, therefore, a

Reale Accademia delle Scienze di Torino **16** (1881): pp. 585–594). Silvio Bedini has recently revived Nelli's faulty claim in "The Instruments of Galileo Galilei," in: *Galileo Man of Science,* ed. Ernan McMullin (New York, Basic Books, 1967), pp. 279–280.

[7] Pp. 42–43, below.

[8] Pp. 43–44, below.

[9] P. 44, below.

[10] P. 38, below.

[11] Pp. 38–39, below.

[12] P. 39.

[13] *De uitvinding der verrekijkers,* pp. 172–173.

[14] Pp. 39–40, below.

[15] P. 47, below.

[16] See, e.g., pp. 44, 46, below.

[17] H. Grotefend, *Zeitrechnung des deutschen mittelalters und der neuzeit* (2 v., Hanover, 1891–1898) **1**: pp. 68, 122, e.g., "Die alte messe ist im Herbst, zwischen den zwein unser Frauentagen assumptio und nativitatis." St. Mary's assumption is celebrated on 15 August, while her birthday falls on 8 September.

[18] In *Ausfuerliche abhandlung von den beruemten zwein reichsmessen so in der reichstadt Frankfurt am Rain jaerlich gehalten werden* (Frankfurt, 1765; no author given), a work

Dutchman was trying to sell a telescope in Frankfurt at just about the same time that Lipperhey set off from Middelburg for The Hague with his instrument.

Cornelis de Waard argued that this Dutch merchant at the Frankfurt fair was the peripatetic Sacharias Janssen.[19] We know that Janssen was a traveling merchant, or peddler, as well as a spectacle-maker. According to de Waard, then, Janssen arrived back in Middelburg around 13 October to find that Lipperhey had gone to The Hague to apply for a patent, and he promptly went to the provincial authorities in Middelburg to show them his telescope. The dates fit rather nicely.

If we agree with de Waard, there were three men who had telescopes by early October, 1608. If we do not wish to identify the man in Frankfurt with the young man in Middelburg (Janssen in both cases, according to de Waard), then there were four: Lipperhey in Middelburg, Metius in Alkmaar, the young man in Middelburg, and the Dutch merchant in Frankfurt. Now Alkmaar is about 75 kilometers north of The Hague and Middelburg about the same distance south of it; Frankfurt is about 500 kilometers to the southeast of the Dutch capital. On the assumption of a single invention just prior to Lipperhey's departure from Middelburg on 25 or 26 September, it becomes rather difficult to account for the prompt reactions of Metius and the young man in Middelburg, as well as the presence of the Dutch merchant with a telescope at the autumn fair in Frankfurt.

In the case of the young man in Middelburg, a plausible explanation would not be difficult to find if we assume Lipperhey to be the first and sole inventor. Presumably this young man heard about the instrument or saw one in Middelburg and copied it, whereupon he presented himself to the provincial authorities. The case of Metius would present some difficulties but these would not be insuperable. The Hague was not only the seat of the States-General, but also of the States of Holland, i.e., the government of the province of Holland. There was close contact between these two bodies. In fact, issues which came before the States-General were often taken up by the States of Holland. On 4 October the latter body discussed Lipperhey's patent application to the States-General and went so far as to appoint a committee to look into this matter further, as is shown by a routine report sent by the delegate of the city of Medemblik (in the northern part of the province Holland) in the States of Holland to his home town. One of the members of this appointed committee was the delegate from the city of Alkmaar.[20] And if the delegate from Medemblik sent routine reports home, we may assume that the delegate of Alkmaar did the same. As Jacob Metius was a member of a prominent family in Alkmaar, this piece of news must have been brought to his attention very quickly upon its arrival there.

Since Metius's patent application was discussed by the States-General on 17 October,[21] he could not have had very much time to act. If we assume that a dispatch left The Hague on the same day as the above-cited dispatch to Medemblik (perhaps in the same post, for the two cities are near each other), on 4 October, and we assume that it took three days to arrive in Alkmaar, then Metius knew about Lipperhey's patent application by the seventh or eighth of October. This means that if he did not have a telescope by that time, he duplicated the invention, built a suitable instrument, and traveled to The Hague with it to submit his own request for a patent, all within ten days (three of which he spent traveling).

This certainly is possible, especially in view of the simplicity of the instrument. But it is not very likely. It would be more reasonable to assume that Jacob Metius already had an instrument (if not a telescope, then at least an instrument which could easily be converted into one) when the report about Lipperhey reached him. At any rate, the States-General thought his request sufficiently legitimate to award him 100 guilders and ask him to improve his instrument further, promising to act on his request after he had submitted an improved instrument.[22] In other words, the States-General treated Metius in exactly the same way as Lipperhey, except that they gave Lipperhey more money.

In trying to account for the Dutch merchant at the Frankfurt fair, under the assumption of a single invention just prior to Lipperhey's patent application, we have a choice. If we identify this merchant with the young man in Middelburg, then presumably this person had the secret before he set out for Frankfurt, sometime towards the end of August. This would mean that the telescope was known by at least two

which discusses the various laws and customs connected with the fairs, there is considerable discussion of regulations which were promulgated periodically to reaffirm the official starting and finishing dates of both the older autumn and the newer spring fair, indicating repeated disregard of the rules in this respect. After citing a number of variations in the dates (pp. 545–546), the anonymous author concludes (p. 546): "Aus welchem iezoangefuerten gar deutlich zu erkennen, dass dieser vor alters uebliche gebrauch, besonders bei der herbstmess, dass sie auf Marienhimmelfart ein- und Mariengeburt ausgelaeutet worden, ongeachtet diese messe, nach obangezogenen ser warscheinlichen gruenden, von gar langen zeiten her, nach leztem festtage, erst ihren anfang genommen und noch iezo nimt, meistens beibehalten worden sei, gleichwie solcher noch, bis auf den heutigen tag, mithin ueber 250. jare, unverrueckt vortwaeret." This means that the autumn fair, in spite of the official regulations, only *began* on or about 8 September and lasted for at least three weeks thereafter.

[19] *De uitvinding der verrekijkers,* pp. 168–170.

[20] Pp. 36–37, below.

[21] P. 40, below.

[22] *Ibid.*

people in Middelburg a whole month before Lipperhey traveled to The Hague with such an instrument, if we want to assume that Lipperhey was the actual inventor. Why would Lipperhey wait so long, especially since the instrument is so simple and easy to copy? If we take the merchant at the Frankfurt fair to be yet a fourth person, then this makes the question as to the early dissemination of the instrument even more difficult. It is unlikely that in this case we are dealing with yet a third native of Middelburg (although it is not impossible), and we have to account for yet a fourth person in possession of a telescope, at the latest simultaneously with Lipperhey's patent application, but 500 kilometers away in this case. Although the individual cases of Metius, the young man in Middelburg, and the merchant at the Frankfurt fair can each be explained in a plausible manner, added together they yield a rather improbable scenario.

De Waard thought he had found a way out of this difficulty. In 1634 Isaac Beeckman, the rector of the Latin School in the city of Dordrecht, wrote in his journal that Johannes Sachariassen, the son of Sacharias Janssen, had told him (presumably during a lesson on lens-grinding) that Sacharias Janssen had made the first telescope in the Netherlands in 1604, copying the instrument from one in the possession of an Italian. This Italian instrument bore the date *anno 190,* presumably to be read *1590.*[23] If this statement is correct, the telescope was present in the Netherlands as early as 1604 but was not invented there.

There are several problems about this statement. Johannes Sachariassen came forward in 1655, when the City Fathers of Middelburg were conducting an official investigation into the origin of the telescope, and claimed that his father had *invented* the telescope in 1590.[24] As de Waard showed, Sacharias Janssen was born around 1588 (at any rate, he married for the first time in 1610).[25] He could thus not have invented the telescope in 1590. Moreover, in 1655 Johannes Sachariassen lied about his own age in order to claim for himself a share in the invention of "long tubes," presumably astronomical telescopes.[26] The 1655 statement is, in fact, a self-serving fabrication, and we can ignore it. If Sacharias Janssen was an unsavory character with scant respect for authority, his son, apparently, was no better in his scant respect for the truth. But why did he pick 1590 as the date of the invention of the telescope when he made his official statement in 1655? Did he simply pick it out of a hat? This would seem strange because Johannes must have known that his father was only about three years old at that date.

If the 1655 statement was a self-serving prevarication, Johannes's 1634 statement, made casually in a conversation with Isaac Beeckman, is to be trusted much more, according to de Waard.[27] It was not made for publication as his later statement was to be—in fact, it did not see the light of day until 1906. It would also explain why in 1655 Johannes picked 1590 as the date of his father's supposed invention. But although de Waard was willing to accept the 1634 statement, recent scholars have been considerably more reluctant to do so, because it presents us with another riddle: why was the instrument a secret between 1590 and 1608? We know that in 1608 the instrument was disseminated with astonishing rapidity, yet we are asked to believe that an Italian who was willing to show it to a spectacle-maker in 1604 had been keeping it a secret for fourteen years prior to that time, and that even after he had shown it to the young Janssen the instrument remained a secret for four more years![28]

The objection is valid and seemingly damning. But suppose we assume that Johannes Sachariassen's 1634 statement was, in fact, more or less correct and that his father did, in fact, copy an instrument from an Italian in 1604. Obviously the Italian did not consider it a particularly precious secret; indeed, it could not have been much of a secret at all. If Johannes was correct, then in 1604 his father copied an instrument consisting of a concave and a convex lens—perhaps in a tube—which magnified slightly. It was probably to be used in the manner indicated by Porta: the correction of defective vision. The reason why it created no stir was that it was still a "feeble thing" and its potential as an instrument for extending normal vision was not apparent to anyone.

But what evidence do we have to support this contention? Let us return to the letter written by Metius to the States-General. Metius wrote that he,

> having busied himself for a period of about two years, during the time left over from his principal occupation, with the investigation of some hidden knowledge which may have been attained by certain ancients through the use of glass, came to the discovery that by means of a certain instrument which he, the petitioner, was using for another purpose or intention, the sight of him who was using the same could be stretched out in such a manner that with it things could be seen very clearly which otherwise, because of the distance and remoteness of the places, could not be seen other than entirely obscurely and without recognition and clarity. Having noticed this, he, the

[23] P. 53, below.

[24] Pp. 55, 57–58, below.

[25] *De uitvinding der verrekijkers,* pp. 117–118, 322–323.

[26] Pp. 55, 57–58, below. Johannes claimed in 1655 that he was fifty-two years old and that he had helped invent "long tubes" in 1618. (I argue that these "long tubes" were not "astronomical" telescopes, in "The 'Astronomical' Telescope, 1611–1650," *Annali dell' Istituto e Museo di Storia della Scienza* 1 [1976]: pp. 20–22. The record of baptisms in Middleburg shows that Johannes Sachariassen was born in 1611 (*De uitvinding der verrekijkers,* p. 323). In 1618 he was, therefore, only seven years old.

[27] *De uitvinding der verrekijkers,* p. 156.

[28] E.g., Stillman Drake, *Galileo Studies* (Ann Arbor, University of Michigan Press, 1970), pp. 155–156.

petitioner, spent his principal time in trying to improve the same, and he finally reached a point where with his instrument he can see things as far away and as clearly as with the instrument which was recently shown to Your Honors by a citizen and spectacle-maker of Middelburg, according to the judgment of His Excellency [Prince Maurice] himself and of others who tested the respective instruments against each other.[29]

This account fits exactly with the above speculations. What *we* might call a telescopic combination of lenses was used by men like Metius for other purposes. Metius realized at some point that this combination could be used to see farther than with the normal naked eye, and set about to improve it for the new purpose. We do not know, of course, whether that realization came to Metius first, or even independently, or was occasioned by the report concerning Lipperhey's "invention." The device Metius was using must have been very imperfect for purposes of seeing far. As he himself states, he spent all his time trying to improve the device once he realized its potential in a new direction. But his account does show that it is entirely justified to assume that devices consisting of a concave and a convex lens, used probably in efforts to correct certain visual defects (devices such as Porta had described in 1589 [30] and Gualterotti was to mention in 1610 [31]) were not unknown in the Netherlands before 1608. If we accept the 1634 statement of Johannes Sachariassen, therefore, we need not conclude that the *telescope* was invented in Italy in 1590 (or before) and that Sacharias Janssen had one by 1604; all we need conclude is that an optical gadget consisting of a concave and a convex lens was present in Italy and was transmitted to the Netherlands several years before Hans Lipperhey petitioned the States-General for a patent on the telescope.

Nor does the link between Italy and the Netherlands present great problems in this respect. The seven northern provinces of the Netherlands, the Dutch Republic, were locked in a protracted struggle with Spain for their independence. The Spanish army in the Netherlands had many Italian soldiers in it, indeed, its commander Ambrogio Spinola (1569–1630) was himself an Italian. Desertion among mercenaries was a common occurrence, and the city records of Middelburg contain a number of references to Italian deserters who had made their way from Flanders to Zeeland, where Middelburg was their most common landing place.[32]

Middelburg also had a glass factory, established in 1581 (the oldest glass factory in the northern provinces), which was the only such establishment licensed in the province of Zeeland and which flourished after the closing of the Schelde. Among its employees at this time several Italians are listed, and from 1605 the factory was actually run by an Italian named Antonio Miotto.[33] The references to the glass factory in the city records of this time are often connected with efforts on the part of establishments in Amsterdam to lure skilled employees away from the factory in Middelburg.[34] We may thus be fairly certain that Italian glass-making techniques had been introduced in the Middelburg factory and that this factory was sufficiently highly regarded for its counterpart in Amsterdam to try to lure its employees away. One of these Italian employees, or perhaps one of the Italian deserters, could easily have been in possession of an optical gadget such as the one described by Porta in his *Magia naturalis* of 1589 and made by Gualterotti for a cavalry soldier in 1598. And it would have been no particular secret. It should also be pointed out, however, that a man such as Metius could easily have made such a device after reading the statement by Porta in *Magia naturalis,* which was a very well-known book.

On the basis of the foregoing we may draw certain cautious conclusions. The telescope was not invented, so to speak, *ex nihilo*. We have the testimonies of Porta, Gualterotti, and Metius that they had been using a device consisting of a concave and a convex lens for other purposes—in the cases of Porta and Gualterotti for the correction of defective vision. Both Gualterotti and Metius testify that their devices were more or less identical with a telescope (or perhaps we should say "spyglass") or could easily be turned into one. Such devices were available in the Netherlands probably well before 1608, and this would explain why a man like Sacharias Janssen could have copied one from an Italian in 1604, without our having to attribute to Janssen or the Italian an actual telescope at that early date. The question, then, as to who invented the telescope boils down to the question: who first realized that such a device could be used for another purpose and set

[29] P. 40, below.

[30] Pp. 34–35, below. See also p. 15, above.

[31] Pp. 45–46, below. See also p. 19, above. There is perhaps another known instance of someone using such a combination for the correction of sight. In his *Magia universalis* (Wuerzburg, 1657–1659) Gaspar Schott writes: "I shall, however, not conceal the fact that Nicolo Cabeo relates that he knew a certain old man, a Jesuit priest, who, many years before, had heard something about an optical tube with two glasses, concave and convex, and made use of it in reciting his canonical hours because he was rather shortsighted, applying the concave close to the eye and the convex close to the book. He had never thought the thing exotic and never revealed it to others, as he considered it a thing not worth to be divulged" (p. 491). Schott found this statement in Cabeo's *Philosophia experimentalis sive commentaria in IV libros Aristotelis meteorologicorum* (Rome, 1644), which I have not been able to consult.

[32] *De uitvinding der verrekijkers,* pp. 319–322.

[33] *Ibid.,* pp. 307–319. See also *Journal tenu par Isaac Beeckman* 3: "Introduction aux notes sur le rodage et le polissage des verres," p. iv.

[34] *Ibid.,* pp. 311–313.

about adapting and improving it in order to obtain the greatest magnification possible?

On the basis of the available evidence this question cannot be answered. We may however be fairly certain that this realization came first in the Netherlands, not long before September, 1608, yet sufficiently long before to allow several others to get wind of the new instrument. Regardless of where and by whom the "invention" was made, it spread from the city of Middelburg, for Middelburg had one of the few glass factories in the Netherlands and Italian glass-making techniques had been introduced there. For the best quality of glass (and this was still a great problem) one went to that city. Because of the simplicity of the telescope, lens-grinders there would quickly fathom the working of the new device, even if they did not actually invent the instrument themselves. It is tempting to conclude that Jacob Metius of Alkmaar first came to the realization. He came from a family well educated in mathematical subjects. His father had been a military engineer under William of Orange and had served as burgomaster of Alkmaar; his brother was professor of mathematics at the Frisian university of Franeker. Jacob himself tells us that he had been looking for "some hidden knowledge which may have been attained by certain ancients through the use of glass." [35] And what better place to start his investigation than in Porta's *Magia naturalis libri xx?* He would naturally have gone to Middelburg to obtain better lenses, and there Lipperhey and Janssen could easily have figured out what he had in mind. But, except for a dubious and obscure passage in Girolamo Sirtori's *Telescopium: sive ars perficiendi,* published ten years after Lipperhey's patent request,[36] there is no evidence to substantiate Metius's claim further.

The issue is clouded not only by the ease with which the instrument could be copied, but also by the fact that several men, upon hearing the news of the new instrument, discovered that they had for all practical purposes been in the possession of such devices but had been using them for other ends, probably at less than their maximum magnifications. These men then either claimed the invention as their own, as Gualterotti [37] and Metius (if he did not actually make the original discovery) did, or denounced the invention as a hoax, as Porta did.[38]

[35] P. 40, below.

[36] Pp. 48, 50, below.

[37] Pp. 45–46, below.

[38] P. 44, below. Later, when the true importance of the new instrument was known, Porta claimed the invention for himself (*Opere di Galileo Galilei,* Edizione Nazionale [20 v., Florence, 1890–1909] **11**: pp. 611–612), saying: "To many men I have shown the telescope . . . . Upon their return to their own countries, they ascribe the invention to themselves." See Edward Rosen, *The Naming of the Telescope,* p. 6. I use Rosen's translation here.

When all is said and done, we are still left with the fact that the earliest undeniable mention of a telescope is to be found in the letter of 25 September, 1608, which Lipperhey carried to The Hague and that Lipperhey was the first to request a patent on the telescope. But to award the honor of the invention to Lipperhey solely on that basis is an exercise in historical positivism.

V. POSTSCRIPT

When Hans Lipperhey arrived in The Hague with his telescope, a few days before the end of September, 1608, that city was the scene of what amounted to (in modern terms) a peace conference. The negotiations between the Dutch and the Spanish for an interruption of the hostilities had been going on for some time. Both sides were divided among themselves, the Dutch having to cope with a sizable faction which favored continuing the war, and the Spanish and Belgian delegation having to contend with the consequently tough Dutch demands which were almost impossible to reconcile with the equally adamant claims of the Spanish king, Philip III. The Dutch had treaties with both the English and the French which would be affected by the outcome of the negotiations, and therefore these countries had sizable delegations in The Hague. The head of the French delegation, Pierre Jeannin (1540–1622), acted as mediator between the two hostile parties, and the eventual success of the negotiations was due in large part to his efforts. Other countries had sent observers as well, with the result that The Hague was crowded with diplomats and their entourages from all over Europe.

It appears that the States-General's efforts to keep secret the invention brought to The Hague by the humble artisan from Middelburg met with little success. In a sheet published in October, 1608, we find the following account of some of the events in The Hague:

> A few days before the departure of Spinola from The Hague, a spectacle-maker from Middelburg, a humble, very religious and God-fearing man, presented to His Excellency [Prince Maurice] certain glasses by means of which one can detect and see distinctly things three or four miles removed from us as if we were seeing them from a hundred paces. From the tower of The Hague, one clearly sees, with the said glasses, the clock of Delft and the windows of the church of Leiden, despite the fact that these cities are distant from The Hague one-and-a-half, and three-and-a-half hours by road, respectively. When the States [-General] heard about them, they asked His Excellency to see them, and he sent them to them, saying that with these glasses they would see the tricks of the enemy. Spinola too saw them with great amazement and said to Prince [Frederick] Henry: "From now on I could no longer be safe, for you will see me from afar." To which the said prince replied: "We shall forbid our men to shoot at you." The master [spectacle] maker of the said glasses was given three hundred guild-

ers, and was promised more for making others, with the command not to teach the said art to anyone.[1]

Although the States-General had forbidden Lipperhey to make telescopes for anyone else and obviously tried to keep the invention a secret, Prince Maurice felt no compulsion in this direction, if the writer of this piece is correct. We know that Spinola, the commander in chief of the Spanish forces, who was in The Hague for the peace negotiations, left that city on 30 September.[2] Maurice thus shared Lipperhey's secret with the commander of the enemy's forces before Lipperhey had had a chance to apply for a patent! In view of the fact that the writer of this newsletter was very well informed with regard to Lipperhey's initial dealings with the States-General, the above account should be taken seriously. It is to be doubted that after having looked through the device Spinola wasted much time, upon his return to Brussels, in having it duplicated by local craftsmen.

Pierre Jeannin, the head of the French delegation in The Hague, heard about the device and approached Lipperhey, asking him to make several telescopes for him. But Lipperhey refused because he had been expressly forbidden to make instruments for others. Jeannin, however, managed before the end of the year to find a French soldier in Prince Maurice's army who had learned the secret, and sent him back to France with letters for King Henry IV and his chief counsellor, the Duke of Sully.[3] Through such diplomatic channels and the above-cited newsletter (which was reprinted in Lyons in November, 1608), the news of the new invention spread through Europe very quickly. The device itself was for sale in the shops of spectacle-makers in Paris by April, 1609;[4] in May we find one in Milan;[5] in August the telescope had reached Venice[6] and Naples.[7]

Obviously, the "secret" of the composition of the new device was not difficult to figure out. The knowledge that it consisted of two lenses in a tube was enough to allow most investigators to duplicate it within a short period of time. Many of these men recognized the potential of the gadget and set out to improve it. But this turned out to be very difficult. Simon Marius found that the local craftsmen in Anspach could not produce lenses of his specifications, and he was no more successful with the lens-grinders in Nuremburg, the center of German instrument-making, known for its spectacle-makers![8] Girolamo Sirtori traveled all over Europe in search of the "secret" of improving the telescope. He even learned to grind and polish lenses himself—all without success.[9] Only a handful of these early investigators managed to improve the telescope significantly. The first, and by far the most important of these, was Galileo.

Galileo heard about the invention some time during the summer of 1609 and quickly figured out how it worked and made himself a spyglass.[10] By 24 August he had improved it sufficiently to be able to present a nine-powered device to the Venetian Senate. This body, already familiar with the three-powered spyglasses for sale in Venice, was sufficiently impressed with the efforts of the lecturer in mathematics at the University of Padua to double his salary and give him a permanent appointment.[11]

But Galileo did not stop here. Throughout the autumn of 1609 he continued to develop his grinding and polishing techniques and was able to make objective lenses with longer and longer focal lengths. By the beginning of 1610 he was in possession of telescopes which magnified 20 and even 30 times.[12] And his efforts were richly rewarded.

It was not a particularly unexpected act to point a telescope to the heavens. The newsletter printed in The Hague in October, 1608, already mentioned that the new device showed stars which are invisible to the naked eye.[13] But except for showing a few additional stars in well-known constellations such as the Pleiades, the initial low-powered spyglasses showed nothing new. Such a device will not even show much more detail in the moon than can be seen with the naked eye. Galileo not only managed to make telescopes with much higher magnifications, he also adapted the device for astronomical use by supplying his objectives with aperture stops.[14] This made it possible to observe the bright celestial bodies without incapacitating interference caused by optical imperfections. In making these two improvements he transformed the telescope from a gadget into a *scientific instrument.* His celestial discoveries, quickly published in his *Sider-*

[1] Pp. 41–42, below.

[2] Jan den Tex, *Oldenbarnevelt* (2 v., Cambridge, Cambridge University Press, 1973) 2: pp. 397–398.

[3] P. 43, below.

[4] Pp. 44, 46, below.

[5] Pp. 48, 50, below.

[6] Lorenzo Pignoria to Paolo Gualdo, 1 August, 1609, *Le opere di Galileo Galilei* **10**: p. 250.

[7] P. 44, below.

[8] Pp. 47–48, below.

[9] Pp. 48–51, below.

[10] Edward Rosen, "When Did Galileo Make his First Telescope?" *Centaurus* 2 (1951): pp. 44–51; Stillman Drake, "Galileo Gleanings VI: Galileo's First Telescopes at Padua and Venice," *Isis* **50** (1959): pp. 245–254; *idem, Galileo Studies: Personality, Tradition, and Revolution* (Ann Arbor, University of Michigan Press, 1970), pp. 140–155.

[11] Pp. 51, 52, below. See also Antonio Favaro, "Galileo Galilei e la Presentazione del Cannocchiale all Repubblica Veneta," *Nuovo archivio veneto* **1**, 1 (1891): pp. 55–75.

[12] Galileo to Belisario Vinta, 7 January, 1610, *Opere* **10**: pp. 273, 277. This is the letter in which Galileo remarks that on that evening he had seen Jupiter accompanied by three *fixed* stars invisible to the naked eye. The following evenings he determined that these stars were not *fixed* but *wandering.*

[13] Pp. 41, 42, below.

[14] *Opere* **10**: p. 278. I thank Stillman Drake for pointing out the importance of Galileo's aperture stops to me.

eus nuncius of March, 1610, shook the world of learning and changed the course of astronomy.

Galileo's haste in publication was not misplaced: others were slowly improving their instruments as well. Thomas Harriot in England was observing the moon, indeed, mapping it, as early as 5 August, 1609 (that is, before Galileo had undertaken serious studies of the heavens with the new device), with a telescope which magnified six times. By 27 July, 1610, he had a 10-powered instrument and by 14 August of that year he was making observations with a 20-powered one.[15] Simon Marius's early telescopic observations are still surrounded by the controversy concerning the date of his first observations of Jupiter's satellites,[16] but we certainly may assume that by the summer of 1610 he had a telescope which could show the satellites of Jupiter.[17] Christoph Scheiner and Nicholas Claude Fabri de Peiresc[18] had good instruments shortly afterwards, although we know almost nothing about them.

Advances also came from another direction. Johannes Kepler, greatly excited by the discoveries made with the new instrument, threw himself into the optical study of lenses and lens systems. Within six months of the publication of *Sidereus nuncius* he had written a treatise on lenses in which he generalized the telescopic effect and outlined other lens combinations which would yield the desired effect.[19] In doing so, he became the inventor of the so-called "astronomical telescope," an instrument which consists of a convex eyepiece and objective. Although Kepler's alternative configuration of lenses was not immediately put into effect, within a decade it had found acceptance in some quarters as the preferred instrument for projecting images of the sun on paper,[20] and by 1630 it was used by some for direct observation of heavenly bodies.[21] By 1650 it had displaced the earlier "Dutch" or "Galilean" instrument in astronomical research.

If Porta was disappointed by the first feeble spyglass that came into his hands in August, 1609, the device was slowly improved and half a century later its useful magnification had been increased by several orders of magnitude. And the aspiration of the *magi* did not die in the seventeenth century although the expressions of it were modulated. If Porta could hope to make a glass "whereby for six hundred miles he [would be able to see] the enemies ships coming,"[22] Robert Hooke could speculate that with sufficient improvement the telescope might some day reveal animals on the moon.[23] We can see our enemies' ships from six

[15] John North, "Thomas Harriot and the First Telescopic Observations of Sunspots," in: John W. Shirley, ed., *Thomas Harriot, Renaissance Scientist* (Oxford, Clarendon Press, 1974), p. 141.

[16] In the preface of his *Mundus jovialis* of 1614, Marius claimed that he had observed the little stars in a straight line with Jupiter for the first time around the end of November, 1609, but that he had not realized for some time that they were satellites. His first notation of the relative positions of the satellites and Jupiter is dated 29 December, 1609, on the Julian calendar, that is, 8 January, 1610, on the Gregorian calendar. Galileo's first notation was made on the previous night.

In *Il saggiatore* of 1623 Galileo accused Marius of having appropriated his observations (*The Controversy on the Comets of 1618*, p. 167). The only serious investigations into this controversy were undertaken in the first decade of the twentieth century. In its program for the year 1900, the Société Hollandaise des Sciences called for a comparative and critical study of the observations of Jupiter's satellites of Galileo and Marius, in order to decide whether or not Galileo's accusation of plagiarism was justified. The only entry was a lengthy study by Joseph Klug in which Galileo's claim was upheld. The referees, however, did not consider Klug's effort worthy of publication in the Society's organ. Antonio Favaro registered his surprise that the prize had not been awarded to the only entry (*Bibliotheca mathematica*, ser. 3, 2 (1901): pp. 220–223). One of the referees, J. A. C. Oudemans, and the Society's secretary, J. Bosscha, then published a lengthy essay, "Galilée et Marius" (*Archives néerlandaises des sciences exactes et naturelles, publiés par la Société Hollandaise des Sciences*, ser. 2, **8** [1903]: pp. 115–189), in which they defended the Society's actions and went on to argue persuasively that Galileo's accusations did not have any serious foundation.

Klug's study, "Simon Marius aus Gunzenhausen und Galileo Galilei," was published several years later in *Abhandelungen der mathematisch-physikalischen klasse der Koeniglich Bayerischen Akademie der Wissenschaften* **22** (1906): pp. 385–526. Bosscha published a critique of Klug's essay the following year ("Simon Marius. Réhabilitation d'un Astronome Calomnié," *Archives néerlandaises*, ser. 2, **12** [1907]: pp. 490–528). If the issue was not settled, these three articles shed an enormous amount of light on Marius's astronomy. A brief review of this episode can be found in J. H. Johnson, "The Discovery of the First Four Satellites of Jupiter," *Jour. British Astronom. Assoc.* **41**, 4 (1931): pp. 164–171.

[17] In a letter to Kepler, written 24 November, 1611 (o.s.), Caspar Odontius reports Simon Marius's observation of the lunar eclipse of 29 December, 1610 (n.s.), and also relates Marius's observation of Jupiter's satellites made the same night. Marius gave careful measurements of the distances of the satellites from Jupiter and gave the periods of the two outermost satellites as 16 days and 10 or 11 days (*Johannes Kepler gesammelte werke* [Munich, C. H. Beck, 1938—] **16**: p. 395). This would indicate that Marius had been observing the satellites for some time.

[18] Scheiner and Cysat were observing through a telescope which magnified "600 or 800 times" (i.e., in area) when they first glimpsed sunspots in March, 1611 (Christoph Scheiner, *Rosa ursina* [Bracciano, 1626–1630], a5^r). Peiresc saw Jupiter's satellites for the first time on 24 November, 1610 (Pierre Humbert, *Un amateur: Peiresc 1580–1637* [Paris, Desclée de Brouwer et Cie, 1933], p. 83).

[19] *Dioptrice* (1611), *Gesammelte werke* **4**: pp. 387–389. *Dioptrice* was composed "inside of a few weeks, in the months of August and September [1610]", see Max Caspar, *Kepler*, tr. C. Doris Hellman (New York, Abelard-Schuman, 1959), p. 198.

[20] Scheiner, *Rosa ursina*, 129^v–130^r.

[21] *Ibid*. See also Francesco Fontana, *Novae coelestium terrestriumque rerum observationes* (Naples, 1646), *passim*.

[22] P. 15, above.

[23] Robert Hooke, *Micrographia* (London, 1665), preface b2^v: "'Tis not unlikely, but that there may be yet invented several other helps for the eye, as much exceeding those already found, as those do the bare eye, such as by which we may perhaps

hundred miles away and we have been to the moon and know that there are no animals on it. If the aspirations of these men were fulfilled by their successors in a manner which was somewhat different from the one they envisaged, they were nevertheless fulfilled. The invention of the telescope, the first extension of one of man's senses, was the first important step on this road.

VI. DOCUMENTS

All dates are given according to the Gregorian calendar. Unless otherwise indicated, the translations are mine.

Roger Bacon, *Epistola de secretis operibus artis et naturae et de nullitate magiae* (*ca.* 1250?). Text taken from the 1618 Hamburg edition, p. 40.

. . . possum enim sic figurari perspicua ut longissime posita appareant propinquissima. & e contrario: ita quod ex incredibili distantia legeremus literas minutissimas, & numeraremus res quantumcunque parvas, & stellas faceremus apparere quo vellemus.

Taken from *Frier Bacon his discovery of the miracles of art, nature, and magick. Faithfully translated out of Dr Dees own copy, by T.M. and never before in English* (London, 1659), p. 20.

Glasses so cast, that things at hand may appear at distance, and things at distance, as hard at hand: yea so farre may the designe be driven, as the least letters may be read, and things reckoned at an incredible distance, yea starres shine in what place you please.

Roger Bacon, *Opus majus* (*ca.* 1267). Text taken from *The "Opus majus" of Roger Bacon edited with introduction and analytical table by John Henry Bridges* (2 v., Oxford, 1897) **2**: pp. 165–166.

De visione fracta majora sunt; nam de facili patet per canones supradictos, quod maxima possunt apparere minima, et e contra, et longe distantia videbuntur propinquissime et e converso. Nam possumus sic figurare perspicua, et taliter ea ordinare respectu nostri visus et rerum, quod frangentur radii et flectentur quorsumcunque voluerimus, ut sub quocunque angulo voluerimus videbimus rem prope vel longe. Et sic ex incredibili distantia legeremus literas minutissimas et pulveres ac arenas numeraremus propter magnitudinem anguli sub quo videremus, et maxima corpora de prope vix videremus propter parvitatem anguli sub quo videremus, nam distantia non facit ad hujusmodi visiones nisi per accidens, sed quantitas anguli. Et sic posset puer apparere gigas, et unus homo videri mons, et in quacunque quantitate, secundum quod possemus hominem videre sub angulo tanto sicut montem, et prope ut volumus. Et sic parvus exercitus videretur maximus, et longe positus apparet prope, et e contra: sic etiam faceremus solem et lunam et stellas descendere secundum apparentiam his inferius, et similiter super capita inimicorum apparere et multa consimilia, ut animus mortalis ignorans veritatem non posset sustinere.

Taken from *The Opus Majus of Roger Bacon; a Translation by Robert Belle Burke* (2 v., Philadelphia, University of Pennsylvania Press, 1928) **2**: p. 582.

The wonders of refracted vision are still greater; for it is easily shown by the rules stated above that very large objects can be made to appear very small, and the reverse, and very distant objects will seem very close at hand, and conversely. For we can so shape transparent bodies, and arrange them in such a way with respect to our sight and objects of vision, that the rays will be refracted and bent in any direction we desire, and under any angle we wish we shall see the object near or at a distance. Thus from an incredible distance we might read the smallest letters and number grains of dust and sand owing to the magnitude of the angle under which we viewed them, and very large bodies close to us we might scarcely see because of the smallness of the angle under which we saw them, for distance in such vision is not a factor except by accident, but the size of the angle is. In this way a child might appear a giant, and a man a mountain. He might appear of any size whatever, as we might see a man under as large an angle as we see a mountain, and close as we wish. Thus a small army might appear very large, and situated at a distance might appear close at hand, and the reverse. So also we might cause the sun, moon, and stars in appearance to descend here below, and similarly to appear above the heads of our enemies, and we might cause many similar phenomena, so that the mind of a man ignorant of the truth could not endure them.

Girolamo Fracastoro, *Homocentrica. Eiusdem de causis criticorum dierum per ea quae in nobis sunt* (Venice, 1538), 18v (Sectio II, cap. 8) and 58r (Sectio III, cap. 23).

F. 18v:

. . . per duo specilla ocularia si quis perspiciat, altero alteri superposito majora multo et propinquiora videbit omnia.

If anyone looks through two spectacle lenses, one placed on top of the other, he will see everything much larger and closer.

F. 58r.

quinimo quaedam specilla ocularia fiunt tantae densitatis, ut si per ea quis aut lunam, aut aliud syderum spectet, adeo propinqua illa iudicet, ut ne turres ipsas excedant. quare nec mirum videri debet si per orbium partes idem quoque contingat.

Indeed, certain spectacle lenses are made of such density, that if someone looks through them at the Moon or at another star, he will judge them to be so close that they do not even appear to exceed the steeples themselves [in height]. This is why one should not be surprised if the same also occurs through the parts of the [heavenly] orbs.

---

be able to discover *living Creatures* in the Moon, or other Planets . . . ."

Robert Recorde, *The pathway to knowledg, containing the first principles of geometrie* (London, 1551), 7th and 8th pages, preface.

But to retourne agayne to Archimedes, he dyd also by arte perspective (whiche is a parte of geometrie) devise such glasses within the towne of Syracusae, that dyd bourne their ennemies shyppes a great way from the towne, whyche was a mervaylous politike thynge. And if I shulde repete the varietees of suche straunge inventions, as Archimedes and others have wrought by geometrie, I should not onely excede the order of a preface, but I should also speake of suche thynges as can not well be understande in talke, without somme knowledge in the principles of geometrie.

But this will I promyse, that if I may perceave my paynes to be thankfully taken, I wyll not onely write of suche pleasant inventions, declaryng what they were, but also wil teache howe a great numbre of them were wroughte, that they may be practised in this tyme also, Wherby shall be plainly perceaved, that many thynges seme impossible to be done, whiche by arte may very well be wrought. And whan they be wrought, and the reason therof not understande, than say the vulgare people, that those thynges are done by negromancy. And hereof came it that fryer Bakon was accompted so greate a negromancier, whiche never used that arte (by any coniecture that I can fynde) but was in geometrie and other mathematicall sciences so experte, that he coulde dooe by theim suche thynges as were wonderfull in the syght of most people.

Great talke there is of a glasse that he made in Oxforde, in whiche men myght see thynges that were doon in other places, and that was judged to be done by power of evyll spirites. But I knowe the reason of it to bee good and naturall, and to be wrought by geometrie (sythe perspective is a parte of it) and to stande as well with reason as to see your face in a common glasse. But this conclusion and other dyvers of lyke sorte, are more mete for princes, for sundry causes, than for other men, and ought not to bee taught commonly. Yet to repete it, I thought good for this cause, that the worthynes of geometry myght the better be knowen, & partly understanding geven, what wonderfull thynges may be wrought by it, and so consequently how pleasant it is, and how necessary also.

John Dee, from *The elements of geometrie of the most auncient philosopher Euclide of Megara. Faithfully (now first) translated into the Englishe toung, by H. Billingsley . . . With a very fruitfull praeface made by M. I. Dee* (London, 1570), preface, a.iiii^{v}–b.i^{r}.

And no small skill ought he to have, that should make true report, or nere the truth, of the numbers and Summes, of footemen or horsemen, in the Enemyes ordring. Afarre of, to make an estimate, betwene nere termes of More and Lesse, is not a thyng very rife, among those that gladly would do it. Great pollicy may be used of the Capitaines, (at tymes fete, and in places convenient) as to use Figures, which make greatest shew, of so many as he hath: and using the advauntage of the three kindes of usuall spaces: (betwene footemen or horsemen) to take the largest: or when he would seme to have few, (beyng many:) contrarywise, in Figure, and space. The Herald, Pursevant, Sergeant Royall, Capitaine, or who soever is carefull to come nere the truth herein, besides the Iudgement of his expert eye, his skill in Ordering *Tacticall,* the helpe of his Geometricall instrument: Ring, or Staffe Astronomicall: (commodiously framed for cariage and use) He may wonderfully helpe him selfe, by perspective Glasses. In which, (I trust) our posterity will prove more skillfull and expert, and to greater purposes, then in these dayes, can (almost) be credited to be possible.

Leonard and Thomas Digges, *A geometrical practise, named Pantometria . . . framed by Leonard Digges Gentleman, lately finished by Thomas Digges his sonne* (London, 1571). *Leonard Digges* (Book I, Ch. 21; G.i^{v}–G.ii^{r}.)

But marveylouse are the conclusions that may be perfourmed by glasses concave and convex of circulare and parabolicall fourmes, using for multiplication of beames sometime the ayde of glasses transparent, whiche by fraction should unite or dissipate the images or figures presented by the reflection of other. By these kinde of glasses or rather frames of them, placed in due angles, ye may not onely set out the proportion of an whole region, yea represent before your eye the lively ymage of every towne, village, &c. and that in as little or great space or place as ye will prescribe, but also augment and dilate any parcell thereof, so that whereas at the firste apparance an whole towne shall present it selfe so small and compacte together that ye shall not discerne any difference of streates, ye may by applycation of glasses in due proportion cause any peculiare house, or roume thereof dilate and shew it selfe in as ample fourme as the whole towne firste appeared, so that ye shall discerne any trifle, or reade any letter lying there open, especially if the sonne beames may come unto it, as playnly as if you wer corporally present, although it be distante from you as farre as eye can discrye: But of these conclusions I minde not here more to intreate, having at large in a volume by it selfe opened the miraculous effectes of perspective glasses.[1]

[1] Neither Leonard nor Thomas Digges ever published such a volume.

Thomas Digges (Preface; A.iii^v^.)

But to leave these celestiall causes and things doone of antiquitie long ago, my father by his continual painfull practises, assisted with demonstrations *Mathematicall,* was able, and sundrie times hath by proportionall Glasses duely situate in convenient angles, not onely discovered things farre off, read letters, numbred peeces of money with the very coyne and superscription thereof, cast by some of his freends of purpose uppon Downes in open fieldes, but also seven myles of declared wat hath beene doon at that instante in private places: He hath also sundrie times by the Sunne beames fired Powder, and dischargde Ordinaunce halfe a myle and more distante, whiche things I am the boulder to reporte, for that there are yet living diverse (of these his dooings) *Oculati testes,* and many other matters farre more straunge and rare which I omitte as impertinente to this place.

William Bourne, *Inventions or devices. Very necessary for all generalles and captaines, or leaders of men, as well by sea as by land* (London, 1578), pp. 96–97.

The 110. Devise

For to see any small thing a great distance of from you, it requireth the ayde of two glasses, and one glasse must be made of purpose, and it may be made in such sort, that you may see a small thing a great distance of, as this, to reade a letter that is set open neare a quarter of a myle from you, and also to see a man foure or five miles from you, or to view a Towne or Castell, or to see any window or such like thing sixe or seaven myles from you. And to declare what manner of glasses that these must bee, the one glasse that must be made of purpose, is like the small burning glasses of that kinde of glasse, and must bee round, and set in a frame as those bee, but that it must bee made very large, of a foote, or 14. or 16. inches broade, and the broader the better: and the propertie of this glasse, is this, if that you doo behold any thing thorow the glasse, then your eye being neare unto it, it sheweth it selfe according unto the thing, but as you doo goe backwardes, the thing sheweth bigger and bigger, untill that the thing shall seeme of a monstrous bignesse: but if that you doo goe to farre backe, then it will debate and be smal, & turne the fashion downewards. But now to use this glasse, to see a small thing a great distance, then doo this, the thing or place that you would view and discerne, set that glasse fast, and the middle of the glasse to stand right with the place assigned, and be sure that it doo not stand oblique or awry by no meanes, and that done, then take a very fayre large looking glasse that is well polished, & set that glasse directly right with the polished side against ye first glasse, to the intent to receive the beame or shadow that commeth thorow the first placed glasse, and set it at such a distance off, that the thing shall marke the beame or shadowe so large, that it may serve your turne, and so by that meanes you shall see in the looking glasse a small thing a great distance, for if that the first placed glasse be well made, and very large, you may descerne and knowe the favour or phisnomie of a man a mile of from you: wherefore in my opinion, this is very necessary in divers respects, as the viewing of an army of men, and such other like causes, which I doo omit, &c.

Thomas Digges, *An arithmeticall militare treatise, named Stratioticos . . . Long since attempted by Leonard Digges Gentleman, augmented, digested, and lately finished, by Thomas Digges, his sonne* (London, 1579), pp. 189–190.

And such was his *Foelicitie* and happie successe, not only in these *Conclusions,* but also in the *Optikes* and *Catoptrikes,* that he was able by *Perspective Glasses* duely scituate upon convenient *Angles,* in such sorte to discover every particularitie in the Countrey rounde aboute, wheresoever the *Sunne* beames mighte pearse: As sithence *Archimedes,* (*Bakon* of *Oxforde* only excepted) I have not read of any in *Action* ever able by meanes natural to performe ye like. Which partly grew by the aide he had by one old written booke of the same *Bakons Experiments,* that by straunge adventure, or rather *Destinie,* came to his hands, though chiefelye by conioyning continual laborious *Practise* with his *Mathematical* Studies.

The which upon this occasion I thought not amisse to rehearse, as well for the knowen *Veritie* of the matter (diverse being yet alive that can of their own sight and knowledge beare faithful witnesse, these *Conclusions* being for pleasure commonlye by him with his friendes practised) as also to animate such *Mathematitians* as enjoye that quiet and rest, my froward *Constellations* have hitherto denyed me, to imploy their studies & travels for *Invention* of these rare serviceable *Secretes.*

William Bourne, *A treatise on the properties and qualities of glasses for optical purposes, according to the making, polishing, and grinding of them* (*ca.* 1585). Text taken from *Rara mathematica,* ed. J. O. Halliwell (London, 1839), pp. 32–47.

EPISTLE DEDICATORY

To the Right Honorable, and hys singuler good Lorde, Sir Wilyam Cicil,[2] Baron of Burghley, Knight of the moste noble order of the garter, Lorde Highe Treasurer of Englande: Mr. of the Courte of Wardens

[2] William Cecil, Lord Burghley (1520–1598).

and Liverys, Chancelour of the University of Cambridge, and one of the Queens Majestie's Honorable Privy Counsell.

Right Honorable, fynding myself moste deepely bounde, unto youre Honour, in dyvers respects: And allso youre Honours moste excellent and worthy skilles, and knowledge in all notable, laudable and noble experiences of learning in all maner of causes: And allso, for that of late youre honour hathe had some conference and speache with mee, as concerning the effects and qualityes of glasses, I have thought yt my duty to furnish your desyer, according unto suche simple skill, as God hathe given me, in these causes, Whiche ys muche inferiour unto the knowledg of those, that ys learned and hathe red suche authors, as have written in those causes, and also have better ability and tyme, to seke the effects, and quality thereof, then I have, eyther elles can, or may, by the meanes of my small ability, and greate charge of children: Whiche (otherwyse) yt ys possible that I shoulde have bene better able to have done a nomber of thinges, that now I must of force leave, that perhapps shoulde have bene. And allso aboute seaven yeares passed, uppon occasyon of a certayne written Booke of myne, which I delivered your honour, Wherin was set downe the nature and qualitye of water: As tuchinge ye sinckinge or swymminge of thinges. In sort youre Honoure had some speeche with mee, as touching measuring the moulde of a shipp. Whiche gave mee occasyon, to wryte a little Boke of Statick.[3] Whiche Booke since that tyme, hath beene profitable, and helpped the capacityes, bothe of some sea men, and allso shipp carpenters. Therfore, I have now written this simple, and breefe note, of the effects, and qualityes of glasses, according unto the several formes, facyons, and makyngs of them, and allso the foylinges of them. That ys to saye, the foyle or using of them, that yow may not looke thorowe the Glass: Whiche causeth the Glass to cast a beame unto your eye, according unto the shape, or forme of any thinge, yt standeth against yt. And allso the polishing and grynding of glasse, which causeth sondry effects: As in ye readinge hereafter dothe appeare the merveylous nature and operation of glasses &c. I humbly desyring your Honour, to take this simple rude matter in good parte: And to accept yt as my good Will, allthoughe that the matter ys of none importance.

By your Honoures, dutyfully to Commande,

W. BOURNE

CHAPTER I.

Introduction

Whereas the eye ys the principall member of the body neyther the Body in respect, coulde not moove any distance, but unto perrill, yf yt were not for the sighte of the Eye, whose quality ys moste wonderfull, and hathe the largest preheminence of all the members of the body. For that the Eye ys able to discerne and see any thinge, how farr down that the distance ys from yow: Yf the thinge bee of magnitude or bignes, correspondent unto the distance. Now, this quality of the sighte of the Eye ys of no quantity or bignes: But onely the quality of the eye ys to see, and begynneth as a poynte, withoute any quantity or bignes. As for example you may knowe yt by this: Pricke a hoale in any thinge, with the poynte of a fine needle, and then holde that unto your Eye, and beholde any thinge thorowe the hole, and you may see a greater thinge, if that yt bee any distance from yow: yow may beholde a whole towne, beeynge a greate distance from you, &c. And for that perspective ys the discerning of any thinge either substancyall or accidentall, accordinge to the bignes, and distance, and hath his boundes, betweene too righte lynes, from a poynte: And so extending infinitely from the sight of the Eye, yt showeth yt self according to the quantity or bignes, correspondent unto the distance. And for that perspective ys muche amplifyed and furdered by the vertue and meanes of Glasses, I do thinke yt good to shewe the property of glasses: And suche, as touchinge the nature and quality of glasses, commonly called Lookinge Glasses. Whiche are those sortes of Glasses, that have ffoyle, layde on the backe syde thereof, that causeth the same glasse to cast from yt a beame or shadowe, accordinge unto the forme of that thinge that standeth ageanst yt shewynge yt unto the sighte of the eye. Whereof there ys three severall sortes, accordinge unto the sundry makinge and polisshinge of these glasses: I do not meane sondry sortes of stuffe, ffor that theare ys some sortes of lookinge glasses that are made of metalles, which are commonly called steele glasses. But I do meane three sondry sorts of forms of making them. As in the one sorte, the beame will shewe ytself accordinge to the bignes, as yt ys: And in the other sorte yt will shewe ytself less, as in the order of the making of yt, the face of a man shall not shewe ytself as broade as the nayle of youre hande. And so accordingly you may make the glasse to shewe the face, at what bignes you lyst, until that yt shall shewe the face at inst [*sic*] the bignes thereof. And the thirde sorte of the making of glasses as you may make the glasse, in sorte, that you will make the face as bigg as the whole glasse howe broade soever that glass ys, yow standing at some one assigned distance.

CHAPTER II

In what forme to make Glasses, for to have yt shewe yt self, according to the bignes of the thinge.

For to have a Glasse (to shewe the beame unto the eye of that biggnes that the thinge ys) That glasse

[3] Probably *The treasure for travellers* (London, 1578). This would date the present tract to 1585.

must bee made flatt and playne; and being well polysshed, and smoothe, and well foyled on the back syde. Then you standinge righte with the middle thereof, you shall receave a beame unto your eye of the trewe forme and shape of your face, or any thinge, that standeth directly right against yt. But yf you do stande any thinge oblique, or awrye, so shall you receave the beame, or see any thinge that maketh the lyke triangle, eyther acute, or sharpp, and obtuse or broade; Accordinge unto the angle: that commeth from youre eye, unto the glasse. So shall you receave, or see that beame, accordinge unto the angle on the other syde, whatsoever that yt bee, howe farr soever that thinge ys in distance from the place, or neare hande.

CHAPTER III

In what forme to make glasses for to have them shewe the forme or facyon of any thinge less in bignes then yt ys.

To have a glasse to make any thinge shewe smaller then yt ys, That must be in the makinge thereof made hillye or bossy outwardes, and to have the ffoyle layde on the hollowe or concave syde, and so yt may bee made, that youre face showe in the beame that commeth to the eye, as small as you lyst, or any syse that ys less than your face at your pleasure. And for to have yt shewe very smalle, then let yt be made half a globe, or boawle, as smalle as a Tennys balle, and so the foylle layde on the concave or hollow syde. So shall the beame that ys cast unto your eye, shewe your whole face, not to be so bigg as the nayle of your fingeres, and so you may make yt to bee hilly, or bossy outewardes, to have any thinge to shewe of what syse, that you lyst. For if that yt bee made as half a boawle or globe, Then of what syde soever you doo stande, you shall see youre owne face, or any thinge that standeth right as you do stande: And to bee in swellenes accordinge to the forme of the hylling or bossing outwardes. And allso as many tymes, that you doo see looking glasses, which make the face longer or broder then the forme or proportyon of your face, the reason thereof ys, that yt hilleth or bosseth more one way, then yt dothe another way. And that way that yt bendeth moste outewardes, that way yt maketh most narrowest, and that way, that the glasse ys most myghtest, that way it sheweth the face most longest. For yf that a glasse were made righte one way, and rounde outwardes the other way, and the foyle layde on the hollow syde, Then that Glasse woulde make the face, the streighte way, the inst [*sic*] lengthe of the face, and the other way narrowe accordinge unto the roundinge of the glasse. So that all sortes of Lookinge glasses, that dothe bosse or hyll owtewardes, dothe shewe the thinge less then yt ys, accordinge unto the bendinge hilling or bossinge outwarde.

CHAPTER IV

In what maner of forme to make Lookinge Glasses, to make any thinge shewe bigger then yt ys.

To make lookinge Glasses for to shewe any thinge bigger then yt ys, That Glasse muste bee made very large: for elles yt will not conteyne any quantitye in sighe; and this glasse must bee Concave inwardes, and well pollyshed of the hollowe or concave syde: and then the foylle must bee layde on that syde that doth swell, as a hyll, and bosse outwarde. And then this glasse, the property of yt ys, to make all thinges which are seene in yt to seem muche bigger then yt ys to the syghte of the Eye, and at some appoynted distance, from the glasse, accordinge to the forme of the hollowness, the thinge will seeme at the biggest, and so yow standinge nearer the thinge will seeme less, unto the sighte of the eye: so that, accordinge unto the forme of the concavity or hollownes, and at some appointed distance from hym that looketh into the glasse, And yf that the glasse were a yearde broade, the beame that shoulde come unto his eye, shall showe his face as broade, as the whole glasse, And to see his face in this glasse, hee must stande righte with the middle of the glasse, &c. And these sortes of glasses ys very necessary for perspective: for that yt maketh a large beame, whereby that a small thinge may be seene, at a greate distance from you: and especially to bee amplified by the ayde of other glasses, &c.

CHAPTER V

In what order to make a glass, that yow may looke thorow, that shall forther your sighte, and to have a small thynge to seem bigg, which ys very necessary for perspective: And yt may bee so made, that you may discerne a small thinge, a greate distance, and specyally by the ayde of other glasses.

And nowe furdermore, as I have shewed before, the forme, and facyon of glasses, that dothe reflect a beame from the glasse, commonly called Lookinge Glasses; So in lyke manner I will shewe you the makinge of Glasses called perspective glasses, that do helpe sighte, by the meanes of the beame, that pearceth commonly thorowe the glasse. And first for makinge of the smallest sorte of them, commonly called spectacle glasses. These sortes of glasses ys grounde uppon a toole of Iron, made of purpose, somewhat hollowe, or concave inwardes. And may be made of any kynde of glasse, but the clearer the better. And so that Glasse, after that yt ys full rounde, ys made fast with syman uppon a smalle blocke, and so grounde by hande, untill that yt ys bothe smoothe and allso thynne, by the edges, or sydes, but thickest in the middle. And then yt ys the quality or property of the Glasse that ys cleare, to shewe all thinge, that ys seene through yt, to seeme bigger and perfecter, then that yow may see yt withoute the Glasse, and the thynner unto the sydes and edges. And the thicker that yt ys in the middle, the bigger or larger any thinge sheweth unto the eye, and yf the glasse bee very cleare, the more perfecter, &c. And now allso in lyke manner for to make a glasse for perspective, for to beholde, and see any thinge, that ys of greate distance from yow, which ys very necessary: for to viewe an army of men, or any castle, or forte, or such other lyke causes. Then they must prepare very cleare, and white Glasse that may bee rounde, and beare a foote in diameter; as fyne and white Vennys Glasse. And the larger, the better: and allso yt must bee of a good thickness, and then yt must bee grounde uppon a toole fitt for the purpose. Beynge sett fyrst uppon a syman block, and firste, grynde on the one syde, and then on ye other syde, untill that the sydes bee very thynn, and the middle thicke. And for that yf the glasse bee very thicke, then yt will hynder the sighte. Therefore yt must bee grounde untill that the myddle thereof bee not above a quarter of an ynche in thickness: and the sydes or edges very thynne, and so polysshed or cleared. And so sette in a frame meete for the purpose for use: so that yt may not be broken. And so this glasse being made in this forme, Then yt wille have three marvellous operacyons, or qualityes, as hereafter you shall see.

CHAPTER VI

The first, and Principall quality of this Glasse, and ys, as touchinge perspectyve.

The quality of the Glasse, (that ys made as before ys rehearsed) ys, that in the beholding any thinge thorowe the glasse, yow standinge neare unto the Glasse, yt will seeme thorow the glasse to bee but little bigger, then the proportions ys of yt: But as yow do stande further, and further from yt, so shall the perspective beame, that commeth through ye glasse, make the thinge to seeme bigger and bigger, untill suche tyme, that the thinge shall seeme shall seeme [*sic*] of a marvellous bignes: Whereby that these sortes of glasses shall muche proffet them, that desyer to beholde those things that ys of great distance from them: And especially yt will be much amplifyed and furdered, by the receavinge of the beame that commeth thorow the glasse, somewhatt concave or hollowe inwardes and well polysshed as I will hereafter furder declare.

CHAPTER VII

The seconde quality of this glass made in the forme before declared.

The quality of this Glass ys, if that the sunne beames do pearce throughe yt, at a certayne quantity of distance, and that yt will burne any thinge, that ys apte for to take fyer: And this burnynge beame, ys somewhat furder from the glasse, then the perspective beame.

CHAPTER VIII

The thyrde of this kynde of Glass, that ys grounde, and made in that forme before declared, ys to reverse, and turne that thyng that yow do beholde, thorowe ye glass, to stande the contrary way.

And yf that yow doo beholde any thinge thorowe this Glasse, and sett the glasse furder from yowe then the burninge beame, and so extendinge after that what distance that yow list, all suche thinges, that yow doo see or beholde, thoroughe the glasse, the toppes ys turned downwardes. Whether that yt bee trees, hilles, shippes on the water, or any other thinge whatsoever that yt be: As yf that yt were people, yow shall see them thoroughe the Glasse, theyre heades downwardes, and theyre feete upwardes, theire righte hande turned to theyre lefte hande, &c. So that this kynde of Glasse beynge thus grounde hathe three marvellous qualityes. For at some assigned, or appoynted distance, accordinge unto the gryndinge of the Glasse, bothe in his diameter, and thicknes in the middle, and thinnes towardes the sydes. That (beholdinge any thinge thorowe the glasse) yt shall make the best perspective beame: So that the thinge that yow doo see thorowe shall seeme very large and greate: and more perfitter withall. And allso standing further from the glasse yow shall discerne nothing thorowe the glasse: But like a myst, or water: And at that distance ys the burninge beame, when that yow do holde yt so that the sunne beames doth pearce thorowe yt. And allso yf that yow do stande further from the glasse, and beholde any thinge thorowe the glasse, Then you shall see yt reversed and turned the contrary way, as before ys declared. So that accordinge unto youre severall standinge, nearer, or furder from the Glasse, beholding any thinge thorowe yt. Suche, yt hathe his perspective beame: and then standinge furder from the glasse, and then all thinges seen thorowe, shall shewe unto the sighte of your eye, cleene turned, and reversed another way, whatsoever that yt bee.

CHAPTER IX

The effects what may bee done with these two last sortes of glasses: The one concave with a foyle, uppon the hylly syde, and the other grounde and pollisshed smoothe, the thickest in the myddle, and thinnest towordes the edges or sydes.

For that the hability of my purse ys not able for to reache, or beare the charges, for to seeke thorowly what may bee done with these two sortes of Glasses, that ys to say, the holowe or concave glasse: and allso that glasse, that ys grounde and polysshed rounde, and thickest on the myddle, and thynnest towardes the sydes or edges, Therefore I can say the lesse unto the matter. For that there ys dyvers in this Lande, that can say and dothe knowe muche more, in these causes, then I: and especially Mr. Dee, and allso Mr. Thomas Digges, for that by theyre Learninge, they have reade and seene many moo [*sic*] auctors in those causes: And allso, theyre ability ys suche, that they may the better mayntayne the charges: And also they have more leysure and better tyme to practyze those matters, which ys not possible for mee, for to knowe in a nombre of causes, that thinge that they doo knowe. But notwithstanding upon the smalle proofe and experyence those that bee but unto small purpose, of the skylles and knowlledge of these causes, yet I am assured that the glasse that ys grounde, beynge of very cleare stuffe, and of a good largenes, and placed so, that the beame dothe come thorowe, and so reseaved into a very large concave lookinge glasse, That yt will shewe the thinge of a marvellous largeness, in manner uncredable to bee beleeved of the common people. Wherefore yt ys to be supposed, and allso, I am of that opinyon, that havinge dyvers, and sondry sortes of these concave lookinge glasses, made of a great largeness, That suche the beame, or forme and facyon of any thinge beeyinge of greate distance, from the place, and so reseaved fyrste into one glasse: and so the beame reseaved into another of these concave glasses: and so reseaved from one glasse into another, beeynge so placed at suche a distance, that every glasse dothe make his largest beame. And so yt ys possible, that yt may bee helpped and furdered the one glasse with the other, as the concave lookinge glasse with the other grounde and polysshed glasse. That yt ys lykely yt ys true to see a smalle thinge, of very greate distance. For that the one glasse dothe rayse and enlarge, the beame of the other so wonderfully. So that those things that Mr. Thomas Digges hathe written that his father hathe done, may bee accomplisshed very well, withowte any dowbte of the matter: But that the greatest impediment ys, that yow can not beholde, and see, but the smaller quantity at a tyme.

Thomas Harriot, *A briefe and true report of the new found land of Virginia* (London, 1588). E.4r.

Most thinges they sawe with us, as Mathematicall instruments, sea compasses, the vertue of the loadstone in drawing yron, a perspective glasse whereby was shewed manie strange sightes, burning glasses, wildefire woorkes, gunnes, bookes, writing and reading, spring clocks that seeme to goe of themselves, and manie other thinges that wee had, were so straunge unto them, and so farre exceeded their capacities to comprehend the reason and meanes how they should be made and done, that they thought they were rather the works of gods then of men, or at the leastwhile they had bin given and taught us of the gods.

Giovanbaptista Della Porta, *Magia naturalis libri XX. Ab ipso authore expurgati, & superaucti, in quibus scientiarum naturalium divitiae & delitiae demonstrantur* (Naples, 1589), Book XVII, Chapter 10, p. 269.

Sed id, quod sequitur longe praestantius vobis cogitandi principium affert, scilicet,

Lente crystallina longinqua proxima videre,

Posito enim oculo in eius centro retro lentem, remotam rem conspicator, nam quae remota fuerint, adeo propinqua videbis, ut quasi ea manu tangere videaris, vestes, colores, hominum vultus, ut valde remotos cognoscas amicos. Idem erit

Lente crystallina epistolam remotam legere.

Nam si eodem loco oculum apposueris, et in debita distantia epistola fuerit, literas adeo magnas videbis, ut perspicue legas. Sed si lentem inclinabis, ut per obliquam epistolam inspicias, literas satis maiusculas videbis, ut etiam per viginti passus remotas leges. Et si lentes multiplicare noveris, non vereor quin per centum passus minimam literam conspiceris, ut ex una in alteram maiores reddantur characteres: debilis visus ex visus qualitate specillis utatur. Qui id recte sciverit accomodare, non parvum nanciscetur secretum. Possumus

Lente crystallina idem perfectius efficere.

Concavae lentes, quae longe sunt clarissime cernere faciunt, convexae propinqua; unde ex visus commoditate his frui poteris. Concavo longe parva vides, sed perspicua, convexo propinqua maiora sed, turbida, si utrunque recte componere noveris, & longinqua & proxima maiora & clara videbis. Non parum multis amicis auxilij praestitimus, qui & longinqua obsoleta, proxima turbida conspiciebant, ut omnia perfectissime contuissent.

Taken from *Natural Magick by John Baptista Porta, a Neapolitane: in twenty books* (London, 1658), p. 368.

But that which follows, will afford you a principle far better for your consideration: Namely,

By a Lenticular Crystal to see things that are far off, as if they were close by.

For setting your eye in the Centre of it behind the Lenticular, you are to look upon a thing afar off, and it will shew so neer, that you will think you touch it with your hand: You shall see the clothes, colours, mens faces, and know your friends a great way from you. It is the same

To read an Epistle a great way off with a Lenticular Crystal.

For if you set your eye in the same place, and the Epistle be at a just distance, the letters will seem so great, that you may read them perfectly. But if you incline the Lenticular to behold the Epistle obliquely, the letters will seem so great, that you may read them above twenty paces off. And if you know how to multiply Lenticulars, I fear not but for a hundred paces you may see the smallest letters, that from one to another the Characters will be made greater: a weak sight must use spectacles fit for it. He that can fit this well, hath gain'd no small secret. We may

Do the same more perfectly with a Lenticular Crystal.

Concave Lenticulars will make one see most clearly things that are afar off; but Convexes, things neer hand; so you may use them as your sight requires. With a Concave you shall see small things afar off, very clearly; with a Convex, things neerer to be greater, but more obscurely: if you know how to fit them both together, you shall see both things afar off, and things neer hand, both greater and clearly. I have much helped some of my friends, who saw things afar off, weakly; and what was neer, confusedly, that they might see all things clearly.

Raffael Gualterotti, *Scherzi degli spirite animali dettati con l'occasione de l'oscurazione de l'anno 1605* (Florence, 1605), p. 26.

Dice Aristotile, che intorno al occhio è mestiero, che sia molto luminoso; e tra l'occhio, e la cosa, che si ha à vedere, sia un corpo diafano, trasparente, ed illuminato, a volere, che si faccia ben l'atto visivo; ma io non so vedere questa necessità; percioche il molto lume intorno al occhio impedisce la vista; come provano quegli, che hanno gl'occhi in fuora: & essendo un lume lontano tre, o quattro miglia nel oscurità de la notte, quanto è maggiore il buio, tanto meglio il lume si vede; si che non pure, non è necessario, ma ne è bisogno, ch'intorno al occhio sia l'Aria molto illuminata; e che tra l'occhio, e la cosa, che si deve rimirare, egli sia di mezzo un corpo trasparente, & illuminato; poiche egli si vede benissimo un picciol lume per le tenebre, e che sia molto lontane; e cosa certa é, che quel picciolissimo lume non si rischiara l'aere intorno, se non per picciolissimo spazio, non pure tra se, e l'occhio. E che più? uno rimirando con un solo occhio, per la buia canna d'una Cerbottana, vede meglio rimirando di giorno, che se per quel buio non havessi a far l'atto visivo; che il molto lume del Aere vicino al occhio impediria, non aiuteria la vista; come mostra l'esperienza; che passando la vista per quella canna arriva al Cielo, e vede le Stelle di giorno, che senza essa canna non vede, se non l'Aria illuminata dal Sole, e cosi meglio si fa l'atto visivo per le tenebre, che per lo corpo illuminato.

Aristotle says that it is necessary that it be very luminous around the eye, and that a diaphanous, transparent, and suitably illuminated body is required between the eye and the object to be perceived, for the act of vision to proceed well. But I do not see this necessity, since the great amount of light around the eye impedes vision. This may be tested by those who look out [at night], for if there be a light three or four miles away in the darkness of the night, the greater the darkness the better the light is seen. Hence the requirements that the air around the eye be highly illuminated, and that between the eye and the object which is to be beheld there be a transparent and illuminated body, are needless. For a small light, even though it is far away, is seen very well through the darkness. And it is certain that that very small light does not light up the surrounding air (except for the smallest space), and surely not [all the space] between it and the eye. And what is more, a person looking with one eye through the dark barrel of a musket sees better, when looking at something in daylight, than if he had not been looking through that darkness. For the great amount of light in the air near the eye would impede, not help vision, as is shown by our experience that vision passing through that barrel and arriving in the sky sees the stars during the day, which without this tube are not seen, only the air, illuminated by the Sun, [being visible]. And so much better does the act of vision proceed through the darkness than through an illuminated body.

Johannes Kepler, *De cometis libelli tres* (1619). Text taken from *Gesammelte werke* **8**: p. 157.

Die 16/26 Septembris feria quarta, Pragae coelo sereno, cum ad spectaculum ignium artificialium noctis hora dimidia supra octavam a meridie in ponte substitissem, finitisque spectaculis intra dimidiam horam, rogante amico, vultum ad stellas convertissem, vidi stellam sub Ursa maiorem caeteris, per perspicilla intuitus, quae aequale caeteris fixis lumen mihi sine perspicillis diffundere videbatur. Caudam ipse nullam vidi, sed rogati caeteri, se videre affirmabant.

On 16/26 September [1607], Wednesday, with the sky in Prague clear, I was positioned on the bridge for the fireworks display at half past eight in the evening, and when the spectacle was finished in less than half an hour, I directed my eyes to the stars, upon being asked by a friend, and looking through glasses I saw a star near the Bear, which was larger than the other stars and which to me, when I put the glasses aside, appeared to pour forth as much light as the other fixed stars. I myself saw no tail, but others who were asked affirmed that they saw one.

Letter from the Committee of Councillors of the States of Zeeland in Middelburg to the Zeeland delegation at the States-General in The Hague, 25 September, 1608. Middelburg, Rijksarchief in Zeeland, MSS "Staten van Zeeland," *1633,* f.31[r].

Edele etc.

Den brenger van dese die verclaert seeckere conste te hebben daer mede men seer verre alle dingen can sien al oft die naer by waeren by middel van gesichten van glasen, dewelcke hy pretendeert dat een niewe inventie is, ende souden deselve gaerne eerst communiceren met Zÿne Excellencie. UE. salhem believen aen

Zÿne Excellentie te addresseren ende naer gelegenheyt ende naerdat UE. de conste bevinden behulpich te wesen. . . . [The rest of the letter concerns other matters of state]
Edele enz. den XXV^{en} Septembris, 1608

Hiermede
Raden

Honored, etc.

The bearer of this, who claims to have a certain device by means of which all things at a very great distance can be seen as if they were nearby, by looking through glasses which he claims to be a new invention, would like to communicate the same first to His Excellency [Prince Maurice]. Your Honor will please recommend him to His Excellency, and, as the occasion arises, be helpful to him, according to what you think of the device. . . .
Honored, etc. the XXVth of September, 1608

Herewith,
Councillors

Minutes of the States-General, 2 October, 1608. The Hague, Algemeen Rijksarchief, MSS. "Staten-Generaal," 33 : f.169^r.

[Jovis den II^{en} Octobris 1608.]

Hans Lipperhey

Opte requeste van Hans Lipperhey geboortich van Wesel wonende tot Middelburch brilmaecker gevonden hebbende seecker instrument om verre te sien gelijck d'Heeren Staten gebleken is, versoeckende aldewyle 'tselve instrument niet en dient gedivulgeert, dat hem gegunt soude wordden octroy, voor den tijt van dertich jaeren, daarby een yegelijck verboden soude wordden, 't voorschreven werck ofte instrument nae te maecken, ofte anderssints hem te accorderen een jaarlijcx pensioen, om 't voorschreven werck alleene te maecken, om ten dienste van den Lande gebruyct te wordden, sonder dat aen eenige uuytheemsche coningen vorsten ofte potentaten te mogen vercoopen, is goetgevonden, datmen eenige uuit dese vergadering sal committeren, omme metten suppt. op sijn inventie te communiceren, ende van denselven te verstaen, oft hy dat niet en soude kunnen gebeteren, sulckx datmen daardoor met twee oogen soude cunnen sien, ende vanden selven te verstaen, waermede dathy te contenteren soude sijn, om het rapport daervan gehoort te adviseren ende geresolveert te wordden, oftmen den suppt. een tractement oft het versochte octroy salhebben te accorderen.

[Thursday the IInd of October, 1608]

Hans Lipperhey

On the request of Hans Lipperhey, born in Wesel, living in Middelburg, spectacle-maker, having discovered a certain instrument for seeing far, as has been shown to the Gentlemen of the States, requesting that, since the instrument ought not to be made generally known, he be granted a patent for thirty years under which everyone would be forbidden to imitate the instrument, or otherwise, that he be granted a yearly pension for making the said instrument solely for the use of the land, without being allowed to sell it to any foreign kings, monarchs, or potentates; it has been approved that a committee consisting of several men of this assembly will be appointed in order to communicate with the petitioner about his invention, and to ascertain from the same whether he could improve it so that one could look through it with both eyes, and to ascertain from the same with what he will be content, and, upon having heard the answers to these questions, to advise [this body], at which time it will be decided whether the petitioner will be granted a salary or the requested patent.

Minutes of the States-General, 4 October, 1608. The Hague, Algemeen Rijksarchief, MSS. "Staten-Generaal," 33, f.171^r.

[Sabati den iiij^{en} Octobris 1608.]

Hans Lipperhey

Is goetgevonden dat boven de communicatie den ii^{en} deses gehouden met Hans Lipperhey geboortich van Wesel gevonden hebbende het instrument omme verre te sien noch uuyt elcke provincie een sal committeren, omme het voorschreven instrument te examineren ende proeven opten toren van Zijn Exc.ies quartier, oft d'inventie ende het werck sulckx is, datmen daervan soude geraecken de vruchten te trecken die men meent, ende in sulcken gevallen metten inventeur te handelen, dat hy aan neme binnen tsjaers te maecken sess sulcke instrumenten van christal de roche (daervooren hy eyscht van elck stuk dusent Guldens) dat hy zynen eysch moderere onderbeloften dat hy zyne inventie niemanden anders teenigen daegen en sal overgeven.

[Saturday the iiijth of October, 1608.]

It has been approved that, in addition to the meeting which took place on the second of this month with Hans Lipperhey, born in Wesel, who discovered the instrument for seeing far, one member from each province shall be appointed to a committee for the purpose of examining and trying the instrument in question on the tower of the quarters of His Excellency [Prince Maurice], in order to determine whether the invention and the workmanship are such that the claimed advantages could be obtained from it, and, if this is the case, to negotiate with the inventor about the fabrication, within one year, of six such instruments made of rock crystal (for which he asks a thousand guilders each), and to ask him to moderate his demand and to promise that he will not divulge his invention to anyone else for some time.

Letter from the representative of the city of Medemblik at the States of Holland to the magistrate of Medemblik, 4 October, 1608. C. J. Gonnet, *Inventaris van he archief Medemblik* (The Hague, 1915), no. 66.

Den iiij octobris 1608 voor noen.

. . . Item oock van een brilman die een instrument geinventeert ende gemaect heeft, daermeede zeer verde

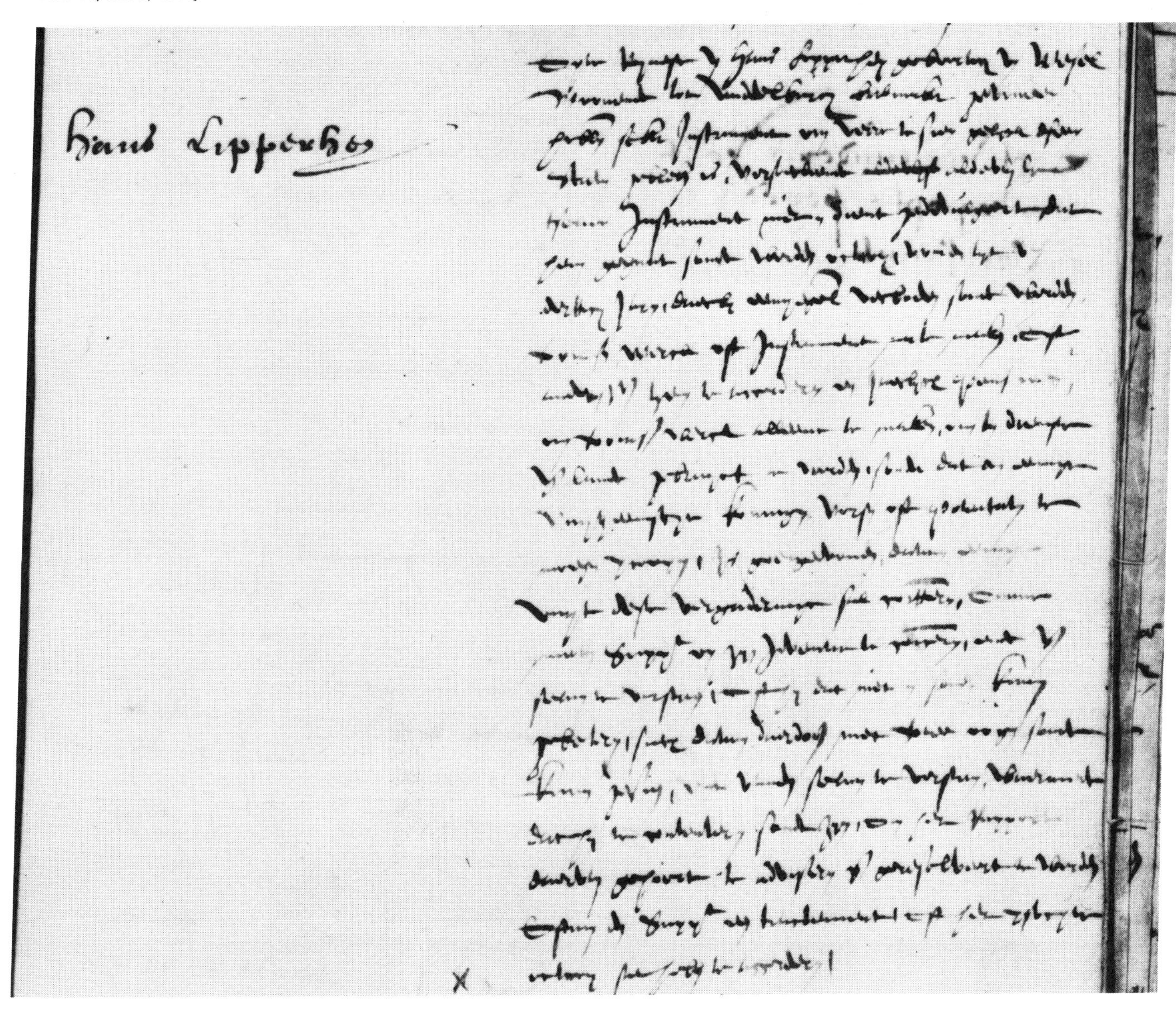

FIG. 2. First entry in the minute book of the States-General concerning Hans Lipperhey's patent application, 2 October, 1608. The Hague, Algemeen Rijksarchief, MSS. "Staten-Generaal," **33**: f. 169^{r}. By courtesy of the Algemeen Rijksarchief.

gesien conde werden, versoeckende octroy om alleene soodanige instrumenten te maecken, met verbot van allen anderen. Worde goetgevonden eenige te committeren die dit instrument noch naerder souden beproeven ende voorts metten inventeur comen in gespreck. Waertoe gedeputeert zijn eene van Dordrecht, van Amsterdam, Rotterdam, Alcmaer, Hoorn, Enchuysen.

The iiijth October, 1608, before noon.
. . . Likewise, in the case of a spectacle-maker who has invented and made an instrument with which one can see very far, asking for a patent allowing him alone to make such instruments and forbidding all others to do so, it was approved that a committee be appointed to investigate this instrument further and also to speak to the inventor, for which purpose one [member] from Dordrecht, Amsterdam, Rotterdam, Alkmaar, Hoorn, and Enkhuizen were appointed.

Account book of the States-General, entry of 5 October, 1608. The Hague, Algemeen Rijksarchief, MSS "Staten-Generaal", *12504,* f.209^{r}.

Die Staten generael etc. ordonneren Philips Doublet ontfanger generael te betalen aen Hans Lipperhey, inventeur van seecker instrument omme verre te sien de somme van drye hondert ponden[4] van veertich grooten[5] tpont ter goeder rekeninge van tgene dat hem by

[4] The pound referred to here is the *Carolusgulden,* i.e., the guilder introduced during the reign of Emperor Charles V.

[5] The Dutch *groot,* as well as the English *groat,* had its origin in the French *gros denier,* a silver coin worth 12 pen-

de Heeren daer toe gedeputeert voor het maken van een sulck instrument ten dienste van tlant toegeleet is. Ende mits, etc. Actum in 'sGravenhage den v^en^ Octobris 1608.

The States-General, etc., order Philips Doublet, receiver-general, to pay to Hans Lipperhey, inventor of a certain instrument for seeing far, the sum of 300 pounds,[4] of forty groats [5] to the pound, towards the sum which the gentlemen commissioned for that purpose have granted him for the making of such an instrument for the use of the land. Etc. Done in The Hague the vth of October, 1608.

Minutes of the States-General, 6 October, 1608. The Hague, Algemeen Rijksarchief, MSS. "Staten-Generaal" **33**, 172^r^.

[Lune den VI^en^ Octobris 1608.]

Johan Lipperhey

Dieheeren Gedeputeerde vande provincien ondersocht hebbende naerder het instrument geinventeert by Johan Lipperhey, bril maecker, ende metten selven gecommuniceert, rapporteren dat zy 't voorschreven instrument naer apparentie bevindende den Lande dienstelijck te sullen vallen, den voorschreven inventeur geboden hebben voor een instrument by hem te maecken van christal de roche voor het Lant drye hondert gulden gereet, ende sess hondert guldens als tselve volmaact ende goed bevonden sal zijn, is goetgevonden ende geresolveert datmen die voorschreven Heeren Gedeputeerde sal authorizeren gelijck deselve geauthorizeert wordden mits desen, omme metten voorschreven Lipperheyt absolutelijck op het maken van tvoorschreven instrument te handelen, ende denselven eenen tijt te limiteren binnen den welcken hy gehouden sal zijn tvoorschreven instrument goet ende wel gemaect te leveren, mits dat dheeren Staten alsdan oock sullen resolveren, ofte den suppt. te accorderen zijn versochte octroy, ofte toeteleggen een jaerlijcx tractement, dies dat hy sal belooven egeen sulcke instrumenten meer te maken sonder consent van heeren Staten.

[Monday the VIth of October, 1608.]

The deputies of the provinces, having more closely examined the instrument invented by Johan Lipperhey, spectacle-maker, and having communicated with the same, report that they find that the instrument in question will apparently be useful to the land, and that they have offered the said inventor, for the making of an instrument of rock crystal for the land, three hundred guilders in advance and six hundred guilders when the same is finished and judged to be good; it has been approved and resolved that these deputies shall be authorized, as they are authorized hereby, to deal bindingly with the said Lipperhey about the fabrication of the said instrument, and to set for the same a time limit within which he will be bound to deliver the said instrument in good and well-made condition, at which time the States will also decide whether to grant the petitioner his requested patent, or to give him a yearly salary, provided that he will promise not to make any more such instruments without the consent of the States.

Minutes of the Committee of Councillors of Zeeland, 14 October, 1608. Middelburg, Rijksarchief in Zeeland, MSS. "Staten van Zeeland," **480**, f lxxvii^v^.

Den xiiii^en^ Octobris 1608.

Is binnen ontboden .... diemen verstaet dat oock de conste soude hebben om instrumenten te maecken om verre dingen nae by te sien, ende is geordonneert daerop te schrÿven aende heeren Gedeputeerde.

The xiiii^th^ October, 1608.

.... was summoned, who, it is understood, also claims to know the art of making instruments for seeing far things near; and it has been ordered that the Deputies [of Zeeland to the States-General] be written about this.

Letter from the Committee of Councillors of Zeeland in Middelburg to the Zeeland delegation at the States-General in The Hague, 14 October, 1608. Middelburg, Rijksarchief in Zeeland, MSS "Staten van Zeeland," **1633**, ff. 32^r^–32^v^.

Edele etc.

Wy hebben U.E. brieven ontfangen ende onder andere daeruut verstaen, t'gene by Zÿne Excellencie ende Heeren Staten-Generael is gedaen ten respecte vande gene die de conste gevonden heeft van verre saecken ende plaetsen als naerby te sien. Wy en hebben daerop nÿet willen naerlaten U.E. te adverteren dat alhier een jongman [is] die oock de conste segt te hebben, ende de selve oock met gelijcke instrument doet blijcken, ende beduchten datter noch meer zijn, ende dat oock andersints nÿet en sal connen secreet blyven want naerdien men weet dat de conste inde warelt is soo sal daer naer getracht worden, besunder naerdien men siet de forme vande buyse, ende daeruut eenichsints de redenen soude connen verstaen om met de gesichten daertoe dienende de conste te vinden, daer aff wÿ U.E. wel hebben willen verwittigen om Zÿne Excellencie ende Heeren Staten Generael daeraff te adverteren, ten eynde wÿ terstont mogen verstaen, t'gene vorder daerin sal behooren gedaen te worden.

Edele enz. den XIIII^en^ Octobris 1608.

Hiermede
Raden

Honored, etc.

We have received the letter from Your Honor and among other things we know from it what has been done by His Excellency [Prince Maurice] and the States-General with respect to the man who has found the art of

---

nies, introduced in 1266. By the beginning of the seventeenth century the Dutch *groot* had the value of half a stiver or one-fortieth of a guilder or pound.

seeing far things and places as if nearby. On this subject we do not want to neglect to inform Your Honor that there is here a young man who says that he also knows the art, and who has demonstrated the same with a similar instrument, and that we believe that there are others as well, and that the art cannot remain secret at any rate, because after it is known that the art exists, attempts will be made to duplicate it, especially after the shape of the tube has been seen, and from it has been surmised to some extent how to go about finding the art with the use of lenses, of which we wish to inform Your Honor, so that you may acquaint His Excellency and the States-General with this, in order that we may quickly be informed as to what further needs to be done in this matter.
Honored, etc. the XIIIIth October, 1608.

Herewith,
Councillors

Letter from Jacob Metius to the States-General, *ca.* 15 October, 1608. *Œuvres complètes de Christiaan Huygens* **13**: pp. 591–593.

Copia copiae exhibitae ab Hadr. Van der Wal 1682. unde patet Jacobum Metium non esse primum inventorem telescopij. sed potius Lippershemium Middelburgensem. [Christiaan Huygens]

Aen de Ed.^{le} Mog. heeren de heeren Staten Generael der Verenighde Nederlanden.

Verthoont me behoorlijcke Eerbiedinge ende reverentie Jacob Adriaenssoon, zoone van Mr Adriaen Anthonissoon oudt borghemeester der Stadt Alkmaer, Uw Ed. Mog. onderdanigen dienaer, hoe dat hij Supp.^{t} omtrent die tijdt van 2 jaeren besigh geweest zijnde, om den tijdt die hem van sijn handwerck ende principaele beroepinghe mochte overigh sijn te emploieren in 't nae soucken van eenige verborgen konsten, die met het gebruijck ende appropieeren van 't glas, bij eenighe ouden te weghe gebracht sijn geweest, gekomen is in ondervindinghe dat bij middel van seecker instrument, 't welck hij Supp.^{t} tot een ander eijnde ofte intentie onder handen was hebbende, 't gesichte van de gheene die 't selve was gebruyckende konde uytstrecken in sulcker voegen dat men daer mede seer bescheydelijck dinghen konde sien, die men anders mids de disstantie ende verheijdt van de plaetse, niet of gansch duysterlijck ende sonder kennisse ofte bescheydt soude konnen sien. 'T welck hy Supp.^{t} vermerckende, heeft hem principaelijck naer dien tijdt geoeffent omme 't selve noch te verbeteren, ende eyntelijck soo verre gebracht, dat men met sijn Instrument een dingh soo verre kan sien ende klaer bekennen als met het Instrument U. Ed. Mog. onlanghs verthoont bij een borgher en Brillemaecker van Middelburgh, volgende het oordeel selve van sijne Exc.^{tie} en andere die de respective Instrumenten tegens malkander hebben geproeft. Niet tegenstaende sijns Supp.^{ts} Instrument maer en is gemaeckt van seer slechte stoffe, ende alleen tot een proeve. 't welck hij Supp.^{t} niet en twijffelt of en soude met verbeteren van de materie oock in 't gebruijck seer gebetert werden, behalve dat hij mede gelooft ende verhoopt metter tijdt de voorschreven inventie in sich selfs soo te verbeteren dat daer noch meerder dienst en vruchten te verwachten sullen sijn. dan alsoo hij Supp.^{t} beducht is dat middelertijdt iemandt hem soude moghen onderwinden de voorschreven sijne instrumenten nae te maecken ofte imiteren, bouwende op de fondamenten die den Supp.^{t} met sijn vernuft, grooten arbeijdt en hooftbrekinghe (door Gods zegeninge) geleijdt heeft, ende hem Supp.^{t} daer mede soude frustreren ende beroven van de vruchten die hij met recht en goede apparentie daer van te verwachten heeft. Soo keert hij Supp.^{t} hem tot U. Ed. Mog. ootmoedelijck versoeckende dat deselve gelieve hem Supp.^{t} te vergunnen Octroy, daer bij eenen iegelijck, de voorschreven inventie voor desen niet gehadt ofte in't werck gestelt hebbende, verboden werde de voorschreven instrumenten in't geheel ofte deel nae te maecken, of die bij soodanighe onvrije personen gemaeckt soude moghen sijn, te koopen of te verkoopen, sonder expres consent van hem Supp.^{t} op de verbeurte van deselve instrumenten, en daer en boven noch van eene somma van hondert car. guld. van 40 grooten 't stuck [6] op elck instrumente contrarie diens gemaeckt, gecoft ofte vercoft, ende dat voor den tijdt van 20 jaeren, of andersins hem Supp.^{t} ten aensiene van de nutheijdt, ende dienste van de voorschreven inventie voor't gemeene Vaderlandt, toe te leggen alsulcke vereeringhe als U. Ed. Mog. nae haere gewoonlijcke goedertierenheijdt en discretie bevinden sullen te behoren, 't welck doende &c. In margine stondt den Supp.^{t} werdt vermaendt voorder te ondersoecken, omme sijne inventie te brenghen totte meeste perfectie, Ende sal alsdan op sijn versochte octroy gedisponeert worden naer behoren. Actum den 17^{n} Oct. 1608. Onder stondt get.^{t}

Aersen.[7]
1608.

Naer gedaene Collatie is dese beneffens de originele requeste met sijn apostille geteeckent als boven, van woorde tot woorde accordeerende bevonden. In Alckmaer den 8 Nov. 1677.

Quod attestor
Joh. H. Metius Not.^{s}
1677

A copy of the copy shown by Adriaan van der Wal in 1682, whence it is evident that Jacob Metius was not the first inventor of the telescope, but rather Lippersheim of Middelburg. [Christiaan Huygens]

To the Honorable Mighty Gentlemen, the Gentlemen States-General of the United Netherlands.

[6] See p. 37, notes 4 and 5.

[7] Cornelis Aerssen(s) (1545–1627) served as the clerk of the States-General from 1584 to 1623.

Jacob Adriaenszoon [Metius], son of Mr. Adriaen Anthoniszoon, ex-burgomaster of the city of Alkmaar, Your Honors' humble servant, testifies that he, the petitioner, having busied himself for a period of about two years, during the time left over from his principal occupation, with the investigation of some hidden knowledge which may have been attained by certain ancients through the use of glass, came to the discovery that by means of a certain instrument which he, the petitioner, was using for another purpose or intention, the sight of him who was using the same could be stretched out in such a manner that with it things could be seen very clearly which otherwise, because of the distance and remoteness of the places, could not be seen other than entirely obscurely and without recognition and clarity. Having noticed this, he, the petitioner, spent his principal time in trying to improve the same, and he finally reached a point where with his instrument he can see things as far away and as clearly as with the instrument which was recently shown to Your Honors by a citizen and spectacle-maker of Middelburg, according to the judgment of His Excellency [Prince Maurice] himself and of others who tested the respective instruments against each other. This is notwithstanding the fact that the instrument of the petitioner was made for the most part of poor materials, and only for a test. And he, the petitioner, does not doubt that with improvement of the materials the instrument would also improve much in use, besides which he also believes and hopes, in time, to improve the said invention itself to such a degree that still more services and fruits can be expected thereof. As he, the petitioner, is afraid that someone may soon imitate the said instrument, building on the foundations which he, the petitioner, has laid by means of his ingenuity, great labor and care (through God's blessings), and that he would thereby be frustrated and robbed of the fruits which he may rightfully, and in all likelihood expect from it, he, the petitioner, therefore turns to Your Honors, humbly requesting that it will please You to grant him, the petitioner, a patent, whereby anyone who did not have the said invention, or was not using it before this, will be forbidden to imitate the said instrument in its entirety or in part, or to buy or sell those which might have been made by unlicensed persons, without the express consent of him, the petitioner, under pains of confiscation of the same instruments and a fine of one hundred Carolus guilders of 40 groats each [6] on every instrument made, bought, or sold contrary to this patent; and that in recognition of the usefulness and the services of the said instrument for the common Fatherland, for the time of twenty years, or some other period of time, Your Honors grant him, the petitioner, such honors as Your Honors' usual good will and discretion will dictate in this matter. Having done which, etc.

[In the margin was written that the petitioner was admonished to investigate further, in order to bring his invention to the greatest perfection, and then his requested patent would be decided on in proper fashion. Enacted the 17th of October, 1608. Signed at the bottom,

Aerssen [7]
1608.]

After collation this has been found to agree word for word with the original request, with its annotation, signed as above. At Alkmaar, 8 November, 1677.

Which I certify
Joh. H. Metius,
Notary
1677.

Minutes of the States-General, 17 October, 1608. The Hague, Algemeen Rijksarchief, MSS. "Staten-Generaal," **33**, f.178^{v}.

[Veneris XVII^{en} Octobris 1608]

Jacob Adriaenszoon Metius

Is Jacob Adriaenszoon, soon van mr. Adriaen Anthoenissen, oudt-burgemeestere der stadt van Alckemaer, versoeckende octroy tot zyne inventie omme het gesichte verre te doen uuytstrecken, in sulcker voegen, datmen daer mede seer bescheydentlijck dingen sal sien die men anders mits de distantie niet oft gansch duysterlijck soude kunnen gesien, toegeleet hondert guldens eens, ende goetgevonden datmen den supp^{t} sal vermanen voorders noch te arbeyden, omme zyne inventie tot meerder perfectie te brengen, als wanneer op zijn versochte octroy gedisponeert sal wordden na behoiren.

[Friday XVIIth October, 1608.]

Jacob Adriaenszoon Metius

To Jacob Adriaenszoon [Metius], son of Mr. Adriaen Anthonissen, ex-burgomaster of the city of Alkmaar, who requests a patent on his invention to stretch out sight in such a manner that, with it, things could be seen very clearly, which otherwise, because of the distance, could not be seen at all, or entirely obscurely, granted one hundred guilders, and approved that the petitioner shall be admonished to work further in order to bring his invention to greater perfection, at which time a decision will be made on his patent in the proper manner.

Account book of the States-General, entry of 17 October, 1608. The Hague, Algemeen Rijksarchief, MSS. "Staten-Generaal," **12504**, f.210^{v}.

Die Staten Generael etc. ordonneren Philips Doublet ontfanger generael te betalen Jacob Adriaansz. de somme van hondert ponden tot XL grooten tpont [8] die hem omme eenige consideratien ende namentlijck tot voerderinge van zijn inventie om verde te sien om deselve in perfectie te brengen toegeleet zijn, Ende mits etc. Gedaen in s'Gravenhage den XVII^{en} Octobris 1608.

The States-General, etc., order Philips Doublet, receiver-general, to pay to Jacob Adriaenszoon [Metius] the sum of one hundred pounds of XL groats to the pound,[8] which have been granted to him for certain considerations, namely, for the furthering of his invention for seeing far, in order to bring the same to perfection. Etc. Done in The Hague the XVIIth of October, 1608.

Ambassades du Roy de Siam [9] *envoyé à l'Excellence du Prince Maurice, arrivé à la Haye le 10 Septemb. 1608.* (The Hague, October, 1608),[10] pp. 9–11.

[8] See p. 37, notes 4 and 5.

[9] This tract was sent to Paolo Sarpi by Francesco Castrino. In a letter to Castrino dated 9 December, 1608, Sarpi ac-

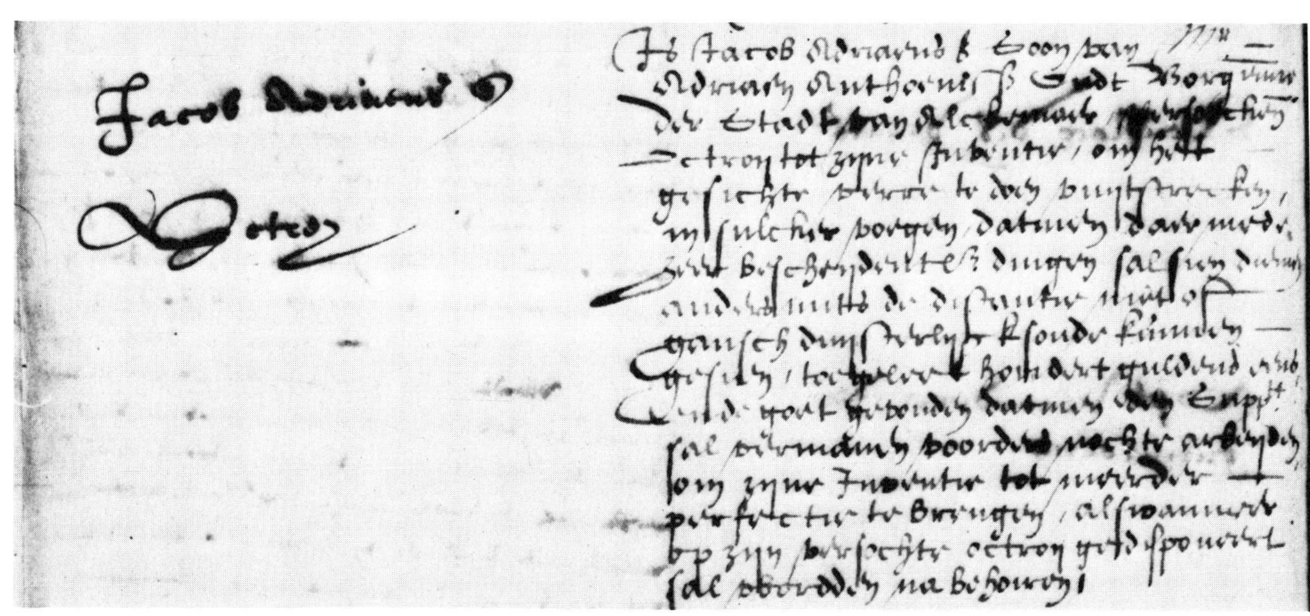

FIG. 3. Entry in the minute book of the States-General concerning the patent application of Jacob Metius, 17 October, 1608. The Hague, Algemeen Rijksarchief, MSS. "Staten-Generaal," 33, f. 178ᵛ. By courtesy of the Algemeen Rijksarchief.

Peu de iours devant le despart de Spinola [11] de la Haye, un faiseur de lunettes de Mildebourg [*sic*] pauvre homme, fort religieux & craignant Dieu, fist present à son Excellence de certaines lunettes, moyennant lesquelles on peut decouvrir & voir distinctement les choses esloignées de nous de trois & quatre lieux, comme si nous les voions à cent pas pres de nous: Estans sur la tour de la Haye on voit par lesdictes lunettes clairement l'horloge de Delft, & les fenestres de l'Eglise de Leyden, nonobstant que lesdites villes soyent esloignées l'une d'une heure & demie, l'autre de trois heures & demi de chemin de la Haye.[12] Messieurs les Estats l'ayant sçeu, envoyerent vers son Excellence pour les voir, qui les leur envoya, disant que par ces lunettes ils verroyent les tromperies de l'ennemi. Spinola aussi les vid avec grand estonnement, et dit à Monsieur le Prince Henry,[13] à ceste heure ie ne scaurois plus estre en seurté, car vous me verrez de loing. A quoy le dit Sieur Prince respondit, nous deffendrons à nos gens de ne tirer point à vous. Le maistre faiseur desdites lunettes a eu trois cent escus, & en aura plus en faisant d'avantage, à la charge de n'apprendre ledit mestier à personne du monde, ce qu'il a promis tresvolontiers, ne voulant point que les ennemis s'en peussent prevaloir contre nous, lesdites lunettes servent fort en des sieges, & en semblables occasions, car d'une lieuë loing & plus, on peut aussi distinctement remarquer toutes choses, comme si elles estoyent tout aupres de nous: & mesmes les estoilles qui ordinairement ne paroissent à nostre veuë & à nos yeux pour leur petitesse & foiblesse de nostre veuë, se peuvent voir par le moyen de cest instrument. Le iour que Spinola partist d'ici, il disna avec son Excell. qui le conduisist demi lieu, & le Prince Henri son frere les accompagna iusques aux navires, ou ils s'embarquerent pour aller à Anvers.

A few days before the departure of Spinola [11] from The Hague, a spectacle-maker from Middelburg, a humble, very religious and God-fearing man, presented to His Excel-

knowledged that he had received a report on the embassy of the king of Siam to Maurice which contained the information about the new glasses. When M. L. Busnelli published this letter in 1927 (*Atti del Reale Istituto Veneto di Scienze, Lettere ed Arti* **87**, 2 (1927): p. 1069) he misread *Siama* and rendered it *Siara,* i.e., *Cearà* a region of northern Brazil. Since no copies of the original editions of this tract were available, Busnelli's example was followed by Stillman Drake (e.g., "Galileo and the Telescope," *Galileo Studies: Personality, Tradition, and Revolution* [Ann Arbor, U. of Michigan Press, 1970], pp. 143, 157 n. 3). But Mr. Jacob Zeitlin has recently found a copy of the original edition which has been published by him in facsimile with an accompanying essay by Stillman Drake. From the original it is clear that the embassay was actually from Siam (*The Unsung Journalist and the Origin of the Telescope* [Los Angeles: Zeitlin & Ver Brugge, 1976], pp. 9–10).

[10] The tract was reprinted in Lyons in November, 1608.

[11] Ambrogio Spinola, Marquess de los Balbases (1569–1630), was a Genoese general in Spanish service. At this time he was the commander in chief of the Spanish forces in the southern Netherlands. He was in The Hague in 1608 to head the Spanish delegation in the negotiations with the Dutch Republic. These negotiations led to the armistice of 1609. He left The Hague on 30 September, 1608.

[12] The distance from The Hague to Delft is 8.6 km; the distance from The Hague to Leiden is 17.6 km.

[13] Frederick Henry (1584–1647) was the youngest son of William the Silent, and the half-brother of Prince Maurice. In 1625 he became commander in chief of the armed forces of the Dutch Republic. Upon Maurice's death the title Prince of Orange reverted to him. From 1625 until his death he was stadholder of Holland, Zeeland, Utrecht, Gelderland, and Overijssel.

lency [Prince Maurice] certain glasses by means of which one can detect and see distinctly things three or four miles removed from us as if we were seeing them from a hundred paces. From the tower of The Hague, one clearly sees, with the said glasses, the clock of Delft and the windows of the church of Leiden, despite the fact that these cities are distant from The Hague one-and-a-half, and three-and-a-half hours by road, respectively.[12] When the States[-General] heard about them, they asked His Excellency to see them, and he sent them to them, saying that with these glasses they would see the tricks of the enemy. Spinola too saw them with great amazement and said to Prince [Frederick] Henry:[13] "From now on I could no longer be safe, for you will see me from afar." To which the said prince replied: "We shall forbid our men to shoot at you." The master [spectacle] maker of the said glasses was given three hundred guilders, and was promised more for making others, with the command not to teach the said art to anyone. This he promised willingly, not wishing that the enemies would be able to avail themselves of them against us. The said glasses are very useful in sieges and similar occasions, for from a mile and more away one can detect all things as distinctly as if they were very close to us. And even the stars which ordinarily are invisible to our sight and our eyes, because of their smallness and the weakness of our sight, can be seen by means of this instrument. The day on which Spinola departed from here, he dined with His Excellency, who accompanied him for half a mile, and Prince [Frederick] Henry, his brother, accompanied them to the ships, where Spinola's party embarked to go to Antwerp.

Minutes of the States-General, 11 December, 1608. The Hague, Algemeen Rijksarchief, MSS. "Staten-Generaal," 33, f.217r.

[Jovis den XIen Decembris 1608.]

Hans Lipperhey

Is gelesen de requeste van Hans Lipperhey, brilmaker ende inventeur van seker instrument om verde te sien, maer niet eyntelijck daerop geresolveert, dan dat eenige zijn gecommitteert om naerder opte voorschreven inventie metten suppliant te spreken, te weeten die Heeren Van Dorth, Magnus, ende Van der Aa.

[Thursday the XIth December, 1608.]

Hans Lipperhey

The request of Hans Lipperhey, spectacle-maker and inventor of a certain instrument for seeing far, was read, but no final decision was made on it, except that several [members] were appointed as a committee to discuss the said invention further with the inventor, namely, the gentlemen Van Dorth, Magnus, and Van der Aa.

Minutes of the States-General, 15 December, 1608. The Hague, Algemeen Rijksarchief, MSS. "Staten-Generaal," 33, f.219v.

[Lunae XVen Decembris 1608.]

Brilleman Lipperhey

De Heeren Magnus ende Oenema in absentie des Heeren Van der Aa ende Boeleszoon rapporteren dat sy gevisiteert hebbende het instrument by den brilleman Lipperhey geïnventeert omme met twee oogen verde te sien, tselve goet hebben gevonden, ende mits dien geproponeert sijnde, oftmen den voorschreven Lipperhey sal accorderen zijn versocht octroy omme voor sekeren tijt tvoorschreven instrument alleene te moegen maecken, ende hem betalen de resterende sesshondert guldens, diehem voor tvoorschreven instrument belooft zijn; is verstaen ende geresolveert, nademael het blijct, dat verscheyden anderen wetenschap hebben van die inventie om verde te sien, datmen het voorschreven versochte octroye des suppliants sal affslaen, maer datmen hem sal lasten binnen sekeren cortten tijt noch twee instrumenten van sijn inventie omme met twee oogen te sien te maecken, ende Dheeren Staten te leveren voordenselven prijs, die hem toegeseet is, ende datmen hem daervoeren alsnoch sal geven ordonnantie van dryehondert guldens, ende vande drye resterende hondert guldens als de voorschreven twee instrumenten als vooren gemaect ende gelevert sullen zijn.

[Monday XVth December, 1608.]

Spectacle-maker Lipperhey

The gentlemen Magnus and Oenema, in the absence of the gentlemen Van der Aa and Boeleszoon, report that they have seen the instrument for seeing far with two eyes, invented by the spectacle-maker Lipperhey, and have found the same to be good. And consequently, the question being raised whether or not to grant the said Lipperhey his requested patent, to be allowed to be the only one to make the instrument for a certain period of time, and whether or not to pay him the remaining six hundred guilders which were promised to him for the said instrument, it is understood and resolved that, since it is evident that several others have knowledge of the invention for seeing far, the requested patent be denied the petitioner, but that he be ordered to make, within a certain short period of time, two more instruments of his invention for seeing with two eyes, and to deliver them to the States for the same price which was accorded him, and that he be given for that purpose a further three hundred guilders, and the remaining three hundred guilders when the two said instruments will have been made as before, and delivered.

Account book of the States-General, entry of 15 December, 1608. The Hague, Algemeen Rijksarchief, MSS. "Staten-Generaal," 12504, f. 213v.

Die Staten generael etc ordonneren Philips Doublet Ontfanger generael te betalen Hans Lipperhey de somme van drye hondert ponden tot XL grooten stuck,[14] op affcortinge vande resterende ses hondert gulden vande negen hondert guldens die hem voor het maecken van seecker instrument om verre te sien, toegeleet sijn. Ende mits etc. Gedaen in 'sGravenhage den vijfthiende december Sesthien hondert ende Acht.

The States-General, etc., order Philips Doublet, receiver-general, to pay to Hans Lipperhey the sum of three

[14] See p. 37, notes 4 and 5.

hundred pounds of XL groats to the pound,[14] as partial payment of the remaining six hundred guilders of the original nine hundred guilders which were granted to him for the fabrication of a certain instrument for seeing far. Etc. Done in The Hague the fifteenth December sixteen hundred and eight.

Pierre Jeannin[15] to King Henry IV, 28 December, 1608. Text taken from *Les negotiations de Monsieur le President Ieannin* (Paris, 1656), 518–519.

. . . Ce porteur qui s'en retourne en France est un soldat de Sedan, lequel a servy quelque temps en la compagnie de Monsieur le Prince Maurice. Il a plusieurs inventions pour la guerre, et sçait faire cette forme de lunettes, trouvée de nouveau en ce païs par un Lunetier de Mildebourg [*sic*], avec lesquelles on voit de fort loing; Les Estats en ont commandé deux pour vostre Majesté à l'ouvrier qui en est l'inventeur. Nous n'eussions emprunté leur faveur pour en avoir, si l'ouvrier en eust voulu faire à nostre priere, mais il l'a refusé, nous disant avoir receu commandement exprés des Estats de n'en faire pour qui que ce soit; nous les luy envoirons à la premiere commodité: & neantmoins ce soldat les fait aussi bien que l'autre, ainsi qu'on le connoist, par l'essay qu'il a fait; aussi n'y a-il pas grande difficulté à imiter cette premiere invention. Nous prions Dieu, SIRE qu'il donne à vostre Majesté en tres-parfaite santé, tres-longue & tresheureuse vie. De la Haye ce vingt-huictième de Decembre 1608. Vos, &c. P. Jeannin, et Russy.

. . . The bearer of this, who is returning to France, is a soldier from Sedan, who served in the army of Prince Maurice for some time. He has several inventions for warfare and knows how to make that form of glasses with which one sees very far, newly found in this country by a spectacle-maker of Middelburg. The States[-General] have ordered two of them for Your Majesty from the artisan who is the inventor of them. We would not have availed ourselves of their favor if the artisan had been willing to make them at our bidding. But he refused, saying that he had been given an express command by the States not to make them for anyone. We shall send them to you at the first opportunity. And at any rate, this soldier makes them just as well as the other man, as is known from the trials he made. Besides, there is no great difficulty in imitating that first invention. We beg God, Sire, to give Your Majesty a very long and very happy life, in perfect health. From The Hague, this twenty-eighth of December, 1608. Yours, etc., P. Jeannin, et Russy.

[15] Pierre Jeannin (1540–1622) was a legal scholar, who became one of the leaders of the League in his important capacity of governor, and president of the parliament of Burgundy. In 1596 he finally made his peace with Henry IV and subsequently occupied many important posts in the French government. In 1608 he was in The Hague to mediate between Spain and the Dutch Republic in the armistice negotiations. The success of these negotiations was largely due to Jeannin's mediation.

Pierre Jeannin to the Duke of Sully,[16] 28 December, 1608. Text taken from *Les negotiations de Monsieur le President* Ieannin (Paris, 1656), 522.

. . . Le porteur de cette lettre est un soldat de Sedan, lequel est de la compagnie de Monsieur le Prince Maurice, tenu fort ingenieux en plusieurs inventions et artifices pour la guerre; Il a aussi depuis peu de iours fait un engin à l'imitation de celuy qui a esté inventé par un Lunetier de Mildebourg [*sic*], pour voir de fort loing. Il vous le fera voir, & vous en fera à l'usage de vostre veuë. J'avois prié le premier inventeur de m'en faire deux, un pour le Roy, & l'autre pour vous: mais les Estats luy ont defendu d'en faire pour qui que ce soit, & les luy ont commandé eux-mesmes pour me les donner, afin que ie vous les envoye, comme ie feray au premier iour, vous supliant tres-humblement que vous me teniez pour ce que ie vous seray perpetuellement, Monsieur, Vostre &c. P. IEANNIN. A la Haye ce vingt-huictième December 1608.

. . . The bearer of this letter is a soldier from Sedan, who comes from the army of Prince Maurice and is held to be very ingenious in several inventions and artifices for warfare. He has also, a few days ago, made a device for seeing very far, in imitation of the one which has been invented by a spectacle-maker of Middelburg. He will show you it, and will make you one suited to your vision. I have asked the first inventor to make me two of them, one for the king and the other for you. But the States [-General] have forbidden him to make them for anyone, and they themselves have ordered them from him, in order to give them to me, so that I can send them to you, which I shall do on the first [possible] day, begging you very humbly to consider me to be what I shall always be to you, Sir. Yours, &c. P. Jeannin. The Hague, this twenty-eighth of December, 1608.

Minutes of the States-General, 13 February, 1609. The Hague, Algemeen Rijksarchief, MSS. "Staten-Generaal," **34**, f.33^r.

[Veneris den XIII^{en} February 1609.]

Hans Lipperhey
Brilmaker

Hans Lipperhey brilmaker heeft gelevert de twee instrumenten omme verre te sien die hem gelast zijn te maken, ende is mitsdien geaccordeert, datmen hem sal depescheren ordonnantie vande drye hondert guldens, die hem alnoch resteren van de negenhondert guldens, die hem voor drye van de voorschreven instrumenten belooft zijn.

[16] Maximilien de Béthune (1560–1641), Baron of Rosny, Duke of Sully, fought in the Protestant army under the Duke of Anjou in the Low Countries, and also fought for that cause in France, under Henry of Navarre, later King Henry IV. When Henry ascended the throne, he placed Béthune (who became the Duke of Sully in 1608) in a number of important posts. At this time he was the king's chief counsellor.

[Friday the XIIIth February, 1609.]

Hans Lipperhey
Spectacle-maker

Hans Lipperhey, spectacle-maker, has delivered the two instruments for seeing far, which he was ordered to make. And it has been approved that he be paid the sum of three hundred guilders which is still due to him of the nine hundred guilders which were promised to him for three of the said instruments.

Account book of the States-General, entry of 13 February, 1609. The Hague, Algemeen Rijksarchief, MSS. "Staten-Generaal," **12504**, f. 218^v.

Die Staten generael etc. ordonneren Philips Doublet ontfanger generael te betalen aen Hans Lipperhey brilmaker de somme van drye hondert ponden van XL grooten tpont,[17] hem resterende aende negen hondert gelijcke ponden hem toegeleet voor het maecken van drye instrumenten omme verre te sien. Ende mits etc. Gadaen den XIII[en] february 1609.

The States-General, etc., order Philips Doublet, receiver-general, to pay to Hans Lipperhey, spectacle-maker, the sum of three hundred pounds of XL groats to the pound,[17] which remained due to him from the nine hundred like pounds granted to him for the fabrication of three instruments for seeing far.

Journal du règne de Henri IV Roi de France et de Navarre. Par Pierre de l'Etoile,[18] *Grand-Audiencier en la Chancellerie de Paris* (4 v., The Hague, 1761) **3**: pp. 513–514, entry of 30 April.

Le jeudi 30 d'Avril, ayant passé sur le pont Marchand, je me suis arrêté chez un Lunetier qui montroit à plusieurs personnes des Lunettes d'une nouvelle invention & usage; ces lunettes sont composées d'un tuyau long d'environ un pied, à chaque bout il y a un verre, mais differens l'un de l'autre; elles servent pour voir distinctement les objets éloignez, qu'on ne voit que très-confusément; on approche cette lunette d'un oeil & on ferme l'autre, & regardant l'objet qu'on veut connoître, il paroit s'approcher, & on le voit distinctement, en sorte qu'on reconnoit une personne de demie-lieuë. On m'a dit qu'on en devoit l'invention à un Lunetier de Midelbourg en Zelande, & que l'année derniere il en avoit fait présent de deux aux Prince Maurice, avec lesquelles on voyoit clairement les objets éloignez de trois ou quatre lieuës: ce Prince les envoya au Conseil des Provinces-unies, qui en recompense donna à l'inventeur trois cent écus, à condition, qu'il n'apprendroit à personne la maniere d'en faire de semblables.

On Thursday 30 April, passing across the Pont Marchand, I stopped at a spectacle-maker's, who was showing glasses of a new invention and use to several people. These glasses are composed of a tube about a foot long, and at each end there is a glass, but [these two glasses are] different from each other. They are used for seeing distant things distinctly, which one sees only very confusedly [with the naked eye]. One approaches these glasses with one eye, and closes the other, and as one looks at the object which one wishes to inspect, it appears to come near, and one sees it distinctly, so that one recognizes a person at half a mile. I have been told that the invention was due to a spectacle-maker of Middelburg in Zeeland, and that last year he presented to Prince Maurice two of them with which things that were three or four miles distant were seen clearly. This prince sent them to the States-General of the United Provinces, who gave the inventor three hundred guilders in compensation, with the condition that he would not show anyone how to make similar ones.

Giovanbaptista Della Porta to Federigo Cesi, 28 August, 1609, *Le opere di Galileo Galilei* (Florence, 1890–1909) **10**: p. 252.

. . . Del secreto dell'occhiale l'ho visto, et è una coglionaria, et è presa dal mio Libro 9 *De refractione;*[19] e la scriverò, che volendola far, V.E. ne harà pur piacere. È un cannelo di stagno di argento, lungo un palmo *ad,* grosso di tre diti di diametro, che ha nel capo *a* un occhiale convesso: vi è un altro canal del medesimo, di 4 diti lungo, che entra nel primo, et ha un concavo nella cima, saldato *b,* come il primo. Mirando con quel solo primo, se vedranno le cose lontane, vicine; ma perchè la vista non si fa nel catheto, paiono oscure et indistinte. Ponendovi dentro l'altro canal concavo, che fa il contrario effetto, se vedranno le cose chiare e dritte: e si entra e cava fuori, come un trombone, sinchè si aggiusti alla vista del riguardante, che tutte son varie.[20] . . .

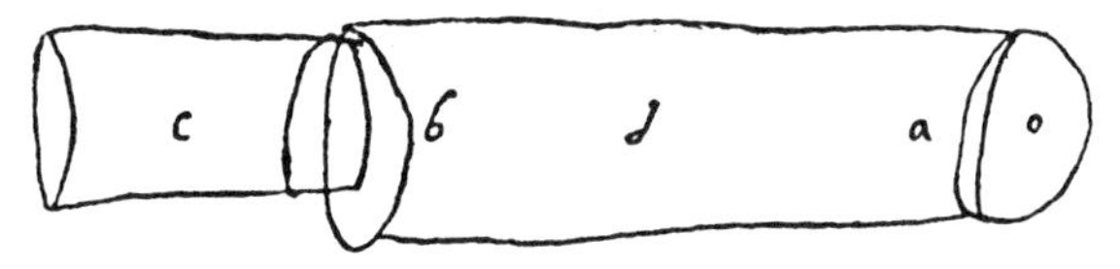

FIG. 4. Sketch made by Giovanbaptista Della Porta of a telescope examined by him in August, 1609. This is the first illustration of a telescope. *Opere di Galileo Galilei* **10**: p. 252.

. . . About the secret of the spectacles, I have seen it, and it is a hoax, and it is taken from the ninth book of my *De refractione.*[19] And I shall describe it, so that if you want to make it, Your Excellency will at least have fun with it. It is a small tube of silvered tin, one palm, *ad,* long, and three inches in diameter, which has a convex glass in the end *a.* There is another tube of the same [material], 4 inches long, which enters into the first one, and in the end *b* it has a concave [glass], which is soldered like the first. If observed with that first one alone, far things are seen near, but because the vision does not occur along the perpendicular, they appear obscure and indistinct. When the other, concave tube is put in, which

[17] See p. 37, notes 4 and 5.

[18] Pierre l'Estoile (1546–1611) came from a family which was prominent in the Paris government. He himself occupied a minor post. He was a collector of rare books, pamphlets, and engravings, and kept a day-by-day journal.

[19] *De refractione optices parte* (Naples, 1953). This book contains nothing which can be interpreted as a reference to the telescope.

[20] This is the first explicit reference to a telescope which can be adjusted to suit the sights of different observers.

gives the contrary effect, things will be seen clear and upright. And it goes in and out like a trombone, so that it adjusts to the visions of the observers, which are all different.[20]

Galileo, *Sidereus nuncius* (March, 1610). Text taken from *Le opere di Galileo Galilei* **3**, i: pp. 60–61.

Mensibus abhinc decem fere, rumor ad aures nostras increpuit, fuisse a quodam Belga Perspicillum elaboratum, cuius beneficio obiecta visibilia, licet ab oculo inspicientis longe dissita, veluti propinqua distincte cernebantur; ac huius profecto admirabilis effectus nonnullae experientiae circumferebantur, quibus fidem alii praebebant, negabant alii. Idem paucos post dies mihi per literas a nobili Gallo Iacobo Badovere ex Lutetia confirmatum est; quod tandem in causa fuit, ut ad rationes inquirendas, necnon media excogitanda, per quae ad consimilis Organi inventionem devenirem, me totum converterem; quam paulo post, doctrinae de refractionibus innixus, assequutus sum: ac tubum primo plumbeum mihi paravi, in cuius extremitatibus vitrea duo Perspicilla, ambo ex altera parte plana, ex altera vero unum sphaerice convexum, alterum vero cavum aptavi; oculum deinde ad cavum admovens obiecta satis magna et propinqua intuitus sum; triplo enim viciniora, nonuplo vero maiora apparebant, quam dum sola naturali acie spectarentur. Alium postmodum exactiorem mihi elaboravi, qui obiecta plusquam sexagesies maiora repraesentabat. Tandem labori nullo nullisque sumptibus parcens, eo a me deventum est, ut Organum mihi construxerim adeo excellens, ut res per ipsum visae millies fere maiores appareant, ac plusquam in terdecupla ratione viciniores, quam si naturali tantum facultate spectentur.

Taken from Stillman Drake, *Discoveries and Opinions of Galileo* (Garden City, Doubleday, 1957), pp. 28–29.

About ten months ago a report reached my ears that a certain Fleming[21] had constructed a spyglass by means of which visible objects, though very distant from the eye of the observer, were distinctly seen as if nearby. Of this truly remarkable effect several experiences were related, to which some persons gave credence while others denied them. A few days later the report was confirmed to me in a letter from a noble Frenchman at Paris, Jacques Badovere, which caused me to apply myself wholeheartedly to inquire into the means by which I might arrive at the invention of a similar instrument. This I did shortly afterwards, my basis being the theory of refraction. First I prepared a tube of lead, at the ends of which I fitted two glass lenses, both plane on one side while on the other side one was spherically convex and the other concave. Then placing my eye near the concave lens I perceived objects satisfactorily large and near, for they appeared three times closer and nine times larger than when seen with the naked eye alone. Next I constructed another one, more accurate, which represented objects as enlarged more than sixty times. Finally, sparing neither labor nor expense, I succeeded in constructing for myself so excellent an instrument that objects seen by means of it appeared nearly one thousand times larger and over thirty times closer than when regarded with our natural vision.

[21] *Belga* should be translated as *Netherlander* or *Dutchman*. See "Note on the word 'Belgium,'" in Pieter Geyl's *The Netherlands in the Seventeenth Century, Part One 1609–1648* (London, Ernest Benn, 1961), pp. 260–262.

Raffael Gualterotti to Galileo, 24 April, 1610. *Le opere di Galileo Galilei* **10**: pp. 341–342.

Molto Ill.re Sig.r

V.S. si partì senza che io potessi dirle alcune cose a bocca di qualche momento; pure forse ritornerà migliore occasione. Fratanto io ho sentito che V.S. ha visto l'occhiale di Mess. Giovambatista milanese, et attribuitoli alcuna loda. Hora, 12 anni sono, io feci uno strumento, ma non già afine di veder gran lontananze e misurar le stelle, ma per benefizio di un cavaliero in giostra e in guerra, e lo proposi al Ser.mo Gran Ferdinando et insieme al' Ill.mo et Excel.mo Sig.r Duca di Bracciano, Don Verginio Orsino; ma parendomi debol cosa, lo trascurai. Pure ancor io, sentendo il romore del Fiammingo, presi i miei vetri e i miei cartoni, e li rimesi insieme, e tornai a considerare il loro uficio, e vedi in terra e 'n cielo molte cose molto meglio che non fa l'occhiale di Giovambatista milanese: e tale strumento mi insegnò fare quel foro che V.S. vide circa a trenta anni sono nela camera mia ala Torre al' Isola, dal qual foro io sino de la mia prima fanciullezza inparai a dubitare del modo del vedere, che la terra refletteva i raggi del sole con gran lume e molto regolatamente, e vi imparai molte bagattelle che io haveva letto esser possibile a farsi, e finalmente lo strumento che 12 anni sono io feci; dal quale indotto, 6 anni sono scrivendo sopra la nuova stella,[22] in proposito del modo del vedere io dissi, che chi voleva veder le stelle di giorno, guatasse per una cerbottana. Hora io ho detto tante parole non per contrariare a la gloria di V.S., ma per esservi a parte molto e molto giustamente, poi che a me si deve quella lode che V.S. dà ad uno Belga, quelo che V.S. può dare ala sua patria. *Mirabil cosa non mi parrà mai Ciò ch'io dirò deli atti fiorentini.* Dio l'ami.

Di Firenze, il di 24 di Aprile 1610.

Di V.S. molto Ill.re Servi.re Aff.mo
Raffael Gualterotti

Very Illustrious Sir,

You left before I had a chance to speak to you about certain things of some importance; but a better occasion will perhaps arise. Meanwhile, I have learned that you have seen the glasses of Mr. Giovambatista of Milan, and have given him some praise. It is now twelve years since I made an instrument, but not for the purpose of seeing

[22] The work referred to here, *Discorso sopra l'apparizione de la nuova stella* (Florence, 1605), does in fact not mention the use of "polar sighting tubes." Gualterotti was mistaken. The passage is in his *Scherzi degli spiriti animali*, published in Florence in the same year, p. 26. See p. 35, above.

great distances and measuring the stars, but rather for the benefit of a cavalry soldier in joust and warfare. And I offered it to the Most Serene Grand Duke Ferdinand, and at the same time to the Most Illustrious and Excellent Lord Duke of Bracciano, Don Verginio Orsino. But as it seemed to me a feeble thing, I neglected it. But hearing the rumor about the Fleming, I again took out my lenses and pasteboard, put them together, and began considering their use. And I saw on Earth and in the sky many things much better than the glasses of Giovambatista of Milan show them. And I found out how to make such an instrument from that hole which you saw about thirty years ago in my room in the tower of Isola, by means of which cavity, even from my earliest childhood, I learned to suspect, about the mode of seeing, that the Earth reflects the rays of the Sun with much light and great regularity. And there I learned many trifles which I had read could be made, and eventually the instrument which I made 12 years ago. And led by this instrument, 6 years ago, when writing about the new star,[22] on the subject of the mode of vision, I stated that he who wanted to see the stars by day should look through a musket barrel. Now I have said so many words not in order to detract from your renown, but in order for you to be well and very justly informed, since to me is due that praise which you give to a Dutchman,[23] and which you can give to your country. *"Astonishing" will never seem to me what I shall [have to] say about the accomplishments of the Florentines.* May God love you.

From Florence, the 24th of April, 1610.
From Your Very Illustrious Sir's
Most Affectionate Servant
Raffael Gualterotti

Le Mercure François, ou la suitte de l'histoire de la paix. Commençant l'an M.DC.V. pour suitte du septenaire du D. Cayer, & finissant au sacre du tres-chrestien roy de France & de Navarre Loys XIII (Paris, 1611), 338ᵛ–339ᵛ.

En ce mesme mois d'Avril [1609] à Paris, il se veit aux boutiques des Lunetiers une nouvelle façon de Lunettes. Aux deux bouts d'un thuyau de fer blanc rond & long d'un pied, il y a deux verrieres, toutes deux dissemblables: Pour regarder ce que l'on veut voir, on ferme un oeil, & à l'autre on en approche la Lunette, avec laquelle on recognoist une personne de demie lieuë: il y a des ouvriers qui en font de meilleures les unes que les autres. Ils disent que ceste invention est venuë de Mildebourg [*sic*] en Zelande, où un Lunetier pauvre homme fit present d'une paire de Lunettes qu'il avoit faictes au Prince Maurice, environ le mois de Septembre de l'an dernier passé, avec lesquelles on voyoit distinctement iusques à trois, & quatre lieuës loing, comme si on eust esté à cent pas prés. Le Prince envoya ces Lunettes au Conseil des Estats durant que l'on traictoit de la Trefve à longues annees avec l' Espagnol & les Archiducs: la lettre qui les accompagnoit portoit, *Par ces Lunettes vous verrez les tromperies de nostre ennemi.* Le prince Henry, frere du Prince Maurice les monstra au Marquis de Spinola, lequel les ayant esprouvees, lui dit, *Ie ne sçaurois plus estre en seureté, car vous me verrez de loing:* & le Prince luy respondit, *Nous defendrons à nos gens de ne point tirer sur vous.* Le conseil des Estats donna trois cents escus à l'inventeur de ces Lunettes, à la charge de n'apprendre à personne du monde son invention; aussi ie pense que celles que l'on vend à Paris, avec lesquelles on ne sçauroit voir une demie lieuë au plus, ne sont comme celles-la de l'ouvrier de Mildebourg [*sic*] : car de la Haye on voyoit clairement l'horloge de Delft, & les fenestres de l'Eglise de Leyden, bien que l'une desdites villes soit esloignee d'une heure & demie de chemin de la Haye, & l'autre de trois. Roger Bachon Anglois, en son Traicté de la merveilleuse puissance de l'art & de la nature,[24] dit, que Cesar du rivage de la Gaule Belgique, front à front de l'Angleterre, avec de certains grands Miroirs ardents, recogneut l'assiette et disposition du camp des Anglois, & de toute la coste de la mer où ils l'attendoient en armes. Beaucoup de belles inventions se sont perduës, mais ce n'est le subjet de notre Histoire de les rapporter icy.[25]

In this same month of April [1609], in Paris, were seen a new form of spectacles in the shops of the spectacle-makers. At the two ends of a round tin tube, a foot long, there are glasses, different from each other. To observe what one wants to see, one closes one eye, and applies the spectacles to the other, through which one recognizes a person from half a mile away. There are some craftsmen who make better ones than others. It is said that this invention came from Middelburg, in Zeeland, where a spectacle-maker, a humble man, presented two of these spectacles which he had made to Prince Maurice, around the month of September of last year [1608]. And with these one clearly saw [things] up to three and four miles away as if they were only a hundred paces away. The prince sent these glasses to the States-General during the negotiations for the long truce with Spain and the archdukes. The letter accompanying them said: "With these glasses you will see the tricks of our enemy." Prince [Frederick] Henry, the brother of Prince Maurice, showed them to the Marquis Spinola, who said to him after trying them: "I could no longer be safe, for you will see me from afar." And the prince replied: "We shall forbid our men to shoot at you." The States-General gave the inventor of these glasses three hundred guilders, with the condition that he would not teach his invention to anyone. I also think that the ones sold in Paris, with which one could see at most half a mile, are not like those of the craftsman from Middelburg. For from The Hague the clock of Delft and the windows of the church in Leiden were seen clearly, although the said cities are one-and-a-half, and three hours by road from The Hague. Roger Bacon, the Englishman, says in his *Epistola de secretis*

[23] See p. 45, note 21, above.

[24] See p. 28, above.

[25] Note the similarities between this account and the accounts of Pierre de l'Estoile (p. 44, above) and of the anonymous author of *Ambassades du Roy de Siam* (pp. 40–42, above). Most likely this latter version was the source of the other two.

operibus[24] that from the coast of Belgian Gaul, opposite England, Caesar inspected the site and disposition of the camp of the English, as well as the entire coast line where they awaited him in arms, by means of burning mirrors. Many wonderful inventions have been lost, but it is not the subject of our history to report them here.[25]

Simon Marius, *Mundus iovialis* (Nuremberg, 1614), preface, the first two unnumbered pages.

Anno 1608. quando celebrabantur Nundinae Francofurtenses Autumnales,[26] versabatur etiam ibidem Nobilissimus, Fortissimus, maximeque strenuus vir, Iohannes Philippus Fuchsius de Bimbach in Möhrn Dominus & Eques Auratus intrepidus belli Dux, &c. Illustrissimorum meorum Principum Consiliarius intimus, totius Matheseos, aliarumque similium scientiarum non saltem fautor & amator, sed & cultor maximus. Inter alia quae tunc ibi gerebantur, accidit, ut Mercator quidam modo nominatum Nobilissimum Virum conveniret, cujus notitiam ante habuerat, & referret quendam Belgam[27] nunc Francofurti esse in nundinis, qui excogitarit instrumentum quoddam, quo mediante, remotissima quaeque obiecta, quasi proxima essent, intueri liceret. Quo cognito multum rogavit dictum Mercatorem, ut belgam[27] illum ad se adduceret, quod tandem obtinuit. Multum igitur disputans cum Belga[27] primo inventore, et de inventi novi veritate nonnihil dubitans Nobilissimus Vir, tandem belga producto instrumento, quod secum attulerat, & cujus alterum vitrum rimam egerat, rei veritatem experiri jussit. Accepto itaque instrumento in manus, & ad objecta directo, ea aliquot vicibus ampliari & multiplicari vidit. Deprehensa itaque veritate instrumenti, quaesivit ex illo, pro quanta pecuniae summa simile instrumentum parare vellet: Belga[27] magnam pecuniae summam poposcit: cum vero intellexerit, quod primum habere non possit, ideo rebus infectis invicem discessum est. Rediens ergo Onoltzbachium dictus Nobilissimus Vir, mihi ad se vocato retulit, excogitatum esse instrumentum, quo remotissima quasi proxima cernerentur. Quae nova ego cum summa admiratione audivi. Cumque hac de re post caenam saepius mecum dissereret, tandem conclusit, necessum scilicet esse ut instrumentum tale duobus constaret vitris, quorum unum esset concavum, alterum vero convexum, & creta accepta proprijs manibus in mensa, quae & qualia intelligeret vitra, delineavit. Accepimus post vitra duo e perspicillis communibus, concavum & convexum, & unum post alterum in conveniente distantia collocavimus, & rei veritatem aliquo modo deprehendimus. Verum cum convexitas vitri ampliantis nimis alta esset, ideo veram convexi vitri figuram gypso impressam Noribergam[28] misit, ad artifices illos, qui perspicilla communia conficiunt, ut similia pararent vitra, at frustra, destituebantur enim instrumentis idoneis, & veram conficiendi rationem illis revelare noluit. Hac ratione nullis interim parcens sumptibus, elapsi sunt menses aliquot. Si modus poliendi vitra nobis cognitus fuisset, statim post reditum a Francofurto, perspicilla optima paravissemus. Interim divulgantur in belgio[27] eiusmodi perspicilla, & transmittitur unum satis bonum, quo valde delectabamur, quod factum est aestate Anni 1609. Ab hoc tempore caepi cum hoc instrumento inspicere coelum & sidera; quando noctu apud saepius memoratum Nobilissimum Virum fui, interdum dabatur mihi potestas portandi domum, praesertim circa finem Novembris, ubi pro more in meo observatorio considerabam astra; . . .

Taken from A. O. Prickard, "The 'Mundus Jovialis' of Simon Marius," *The Observatory* **39** (1916): pp. 367–381, 403–412, 443–452, 498–503; this passage is on pp. 370–371.

In the year 1608, when the Frankfurt autumn fair was going on,[26] it happened that there was at the same place the most noble, gallant, and energetic John Philip Fuchs, of Bimbach in Möhr, Lord and Knight, and a dauntless General, Privy Councillor of my own most illustrious Princes, not only a patron and lover but also an eminent student of all Mathematics and other kindred sciences. Various things went on there, and among others it chanced that a certain merchant met the nobleman mentioned above, whose acquaintance he had formerly made, and told him that there was then present in Frankfurt at the fair a Dutchman,[27] who had invented an instrument by means of which the most distant objects might be seen as though quite near. Hearing this, he begged the merchant to bring the Dutchman[27] to him, which the merchant at last consented to do. Our nobleman had a long discussion with the Dutch[27] first inventor, and felt doubts as to the reality of the new invention. At last the Dutchman[27] produced the instrument, which he had brought with him, and one glass of which was cracked, and told him to make a trial of the truth of his statement. So he took the instrument into his hands, and saw that objects on which it was pointed were magnified several times. Satisfied of the reality of the instrument, he asked the man for what sum he would produce one like it. The Dutchman[27] demanded a large price, and when he understood that he could not get what he first asked, they parted without coming to terms. When he returned to Ansbach, the Nobleman sent for me, and told me that an instrument had been devised by which very remote objects were seen as though quite near. I heard the news with the utmost surprise. He frequently talked the matter over with me after supper, and at last came to the conclusion that such an instrument must necessarily be composed of glasses, of which one was concave and the other convex. He took up a piece of chalk and with his own hand drew a sketch on the table to show what sort of glasses he meant. We afterwards took glasses out of common spectacles, a concave and a convex, and arranged them one behind the other at a convenient distance, and to a certain extent ascertained the truth of the matter. But as the convexity of the magnifying-glass was too great, he made a correct mould

[26] For the dates of this event, see p. 21, above.

[27] Prickard has translated *belga* and *Belgium* as *Belgian* and *Belgium.* I have translated these words as *Dutchman* and *the Netherlands;* see p. 45, note 21, above.

[28] Nuremburg was one of the earliest centers of lens-grinding in Europe. At this time it was still the foremost center of spectacle-makers in Germany.

in plaster of the convex glass, and sent it to Nuremberg[28] to the makers of ordinary spectacles that they might prepare him glasses like it; but it was no good, as they had no suitable tools, and he was unwilling to reveal to them the true principles of the process. No expense was spared, and several months elapsed. If we had been acquainted with the method of polishing glasses, we should have produced excellent spy-glasses immediately after [Fuchs's][29] return from Frankfurt. In the meantime, glasses of the same kind were becoming common in the Netherlands,[27] and a fairly good one was sent, with which we were highly pleased. This was in the summer of 1609. From this time I began to look into the heavens and the stars with this instrument, whenever I was at the house of the nobleman so often mentioned, at night time; sometimes he used to allow me to carry it home, and in particular about the end of November, when I was observing the stars according to my custom in my own observatory.

Adriaen Metius, *Nieuwe geographische onderwysinghe* (Franeker, 1614), p. 15.

Soo wanneer mijn Broeder ghelieven sal zijne ghevondene perspicillen (die bij hem alsnoch rusten) aen den dach te brenghen, soo sal men op dese manier de longitudines der landen perfectelicker connen afmeten, want men door dieselve perspicillen in de Mane zekere hoochten en dalen kan aenschouwen, die onbeweechlijck altijt hare plaetse houden, van welcke men de distantie der sterren tot op een secunde door behulp derselver perspicillen connen afmeten.

When it will please my brother to reveal his invented glasses (which he is still keeping secret), the longitudes of countries can be measured more perfectly. For through these same glasses certain heights and valleys can be seen in the Moon, which always stay in precisely the same places, and distances to stars from these places can be measured to a second with the help of the same glasses.

Adriaen Metius, *Institutiones astronomicae et geographicae, Fondamentale ende grondelijcke onderwysinge van de sterre-konst* (Franeker, 1614), pp. 3–4.

Noch openbaren hen des daghes nevens de sonne veel andere verscheydene planeten, dewelcke by ghene Autoren zijn bekent gheweest, dan werden alleene ghesien door de verre ghesichten, die by mijn Broeder Jacob Adriaensz. over ontrent 6 jaren ghevonden zijn geweest.

These days, besides the Sun, many other planets are revealed which were not known to any authors until they were seen through the far sights, which were found by my brother Jacob Adriaenszoon [Metius], about 6 years ago.

Girolamo Sirtori, *Telescopium: sive ars perficiendi* (Frankfurt, 1618), pp. 23–30.

Prodiit anno 1609. [*sic*] seu Genius, seu alter, vir adhuc ignotus Hollandi specie, qui Midelburgi in Zelandia convenit Ioannem Lippersein, is est vir solo aspectu insigne aliquod praeferens, & perspicillorum artifex; Nemo alter est in ea urbe, & iussit perspicilla plura tam cava quam convexa confici: Condicto die rediit absolutum opus cupiens, atque ut statim habuit prae manibus, bina suscipiens, cavum scilicet & convexum, unum & alterum oculo admovebat & sensim dimovebat, sive ut punctum concursus, sive ut artificis opus probaret, postea soluto artifice abiit. Artifex ingenii minime expers, & novitatis curiosus, coepit idem facere & imitari, nec tarde natura suggessit tubo haec perspicilla condenda, ubi unum absolvit, advolavit in Aulam Principis Mauritij & adinventum obtulit. Princeps habuerit prius, nec ne, suspicandum erat rem militiae utilem, & pernecessariam inter arcana custodiri. Verum ut casu senserit evulgatam, dissimulaverit, industriam, & benevolentiam artificis gratificans. Inde tantae rei novitas per totum effunditur orbem, & plura alia confinguntur spicilla, sed nullum illi contigit melius aut aptius priore (ego vidi & tractavi) adeo ut dicas non Artes solum, sed ipsam Naturam omnia conferre, ut magnis Principibus inserviant. Ferebatur etiam nil praeterea esse hoc adinventum, quam duo spicilla tubo apposita: Et cum Porta in sua Magia de hac re, licet obscure, verba fecisset, & ore tenus etiam cum multis me praesente, videbatur pluribus inesse hanc conceptionem, adeo ut re audita quilibet ingeniosus coeperit sine exemplo pertentare opus. Alij lucri cupiditate Belgae, Galli, Itali quocunque procurrebant, nemo erat qui authorem se non faceret. Mediolanum mense Maio advolavit Gallus qui huiusmodi Telescopium obtulit Comiti de Fuentes, is se socium Hollandi authoris aiebat, Comes cum dedisset Argentario ut tubo argenteo includeret, incidit in manus meas, tractavi, examinavi, & similia confeci, in quibus cum observassem multi ex vitro accidere incommoda contuli me Venetias ut ex opificibus copiam compararem & adhuc artis omnino rudis cuidam tradito spicillo undequaque absoluto ut similia conficeret, nonnihil paecuniae inutiliter prodegi, ac spicillum ammisi nil praeterea edoctus, quam sorte, & laborioso spicillorum delectu rem perficiendam esse. Forte cum unum parassem, imprudens conscenderam Divi Marci Turrim ut eminus experimentum caperem; aliquis e foro novitate prospecta, alios monuit, inde nobilis iuventutis turba tanta curiositate sursum deferebatur, ut parum abfuerit quin me obrueret, modeste tamen, atque humaniter rogato Telescopio coeperunt prospicere, alter alteri tradens: duabus ferme horis hac mora, & inexpectato casu fatigatus, tandem ieiunus stomachus unumquemque domum suam revocans, coepit multitudo rarescere, & ego respirare. Sequenti die memor pridiani periculi, & timens idem futurum si rescirent diversorium de quo abeuntes sollicite percontabantur, Valedixi. Impatienter tamen hoc unum ferebam artem adhuc in incerto habere, & tanto labore parandam, & cogitabam qua ratione possem assequi. Interim fama in maius credebatur, & circumferebantur

[29] Prickard has "immediately after *our* return" (my italics). This is an error.

mendacia plura, quae desiderium augerent, in Belgio, in Hispania huismodi Telescopia reperiri, quae ad tria milliaria dignoscerent hominem, mercatores literis testari. Ego in Hispaniam iter suscepi ratus singularia quaeque certius & citius ibi adfutura. Gerundam cum pervenissem exploravit aliquis me huiusmodi spicillum habere quale per omnium ora ferebatur: Mox adfuit architectus quidam curiosus rogans si posset meum videre Telescopium; Ego aversatus hominis importunitatem, coepi renuere: ille rursus urgere, nec secedere a latere, ita ut in suspicionem venirem hominem utique arti dedictum esse, nec fefellit, nam cum arborem remotam ad satietatem diu esset conspicatus iterum rogavit ut permitterem scrutari, educere, & tractare spicilla, annui, gnarus illum impar aetati onus subire si vellet imitari: Posteaquam vitra tractasset, & diligenter considerasset, duxit me in illius hospitium, & recluso conclavi, reseravit ferramenta artis rubigine consumpta. Is fuerat aliquando perspicillorum artifex, & tota ars ibi latitabat. Ut me sensi Genij artis favore eo perductum, totum me dedi in illius amicitiam, & in illum liberius secretum effudi. Ipse praeterea formas artis libro delineatas ostendit, & roganti permisit ut proportiones tribus tantum punctis exscriberem: Non fuit mihi postea difficile integras assumere, & deinde re diligenter examinata, & cottidie experimentis labore, sumptibus aucta, & confirmata, perficere, & in eam redigere Tabulam, quam tibi patefacio. Noster architectus, ut postea intellexi Frater erat Rogeti Burgundi Barcinonae quondam accolae magnae industriae viri qui artem in Hispaniam primus induxit & stabilivit. Is tres filios suscepit quorum unus literis & Religioni deditus Divi Dominici coetui se addixit: artem ipse monachus delineaverat: Nullibi haec ars exactior quam apud istos fratres Rogetos. Iam videbar artem didicisse qui formas tantum nactus eram, sed tam ex voto mihi cesserat, ut sperarem brevi posse Telescopium perficere, itaque in aulam Regiam properans privati negotii causa, & inde cito expeditus, in patriam redii, ubi ferramenta non nulla conficienda curavi, conductisque mercenariis, coepi observare quae ad artem facerent, & manum paullatim exercere; post diuturnum opus & laborem, cum experimentum caperem perspicillorum, illud imprimis se mihi obiecit, in eadem forma perspicilla fere omnia inter se inaequalia evasisse, cuius rei causam cum diligenter examinarem, comperi ex inaequalitate formae praecipue, deinde ex crassitie vitri inaequali, postremo ex manus ambitu provenire. Mea spe, & labore adhuc frustratus, Romam concessi, quo intellexeram omnium artium copiam affluere, & praeterea multos insignes & praeclaros viros huic studio deditos esse; Nil mihi felicius contingere poterat. Aderat n. Galilaeus cum suo, nunquam interiturae memoriae, Telescopio: Forte quadam die Federicus Princeps Caesius, & Marchio Monticellorum, vir eruditus, & literarum Maecenas invitaverat illum ad coenam, In Vinea quae dicitur Malvasia, ac praterea nonnullos alios literatos;[30] Ante occasum solis cum eo pervenissent, coeperunt Telescopio prospectare Inscriptionem Sixti V. Pontif. in supercilio Ianuae Lateranensis quae distat uno ferme milliari, succedi ego & vidi, & ad satietatem legi inscriptionem. Noctu deinde, & post coenam Iovem & comitantium stellarum motus observavimus, ubi satis recreati tanti luminis adspectu, atque rei curiositate, secesserunt Telescopium scrutaturi, & ipse Galilaeus ut curiositati satisfaceret, eduxit lentem, & cavum spicillum & palam ostendit: Ego interim tubum scrutatus, atque dimensus, lentem quoque deinde tractavi & consideravi, adeo ut possim ex fide, ex arte atque experientia referre qualis sit: Id unum mihi deerat, exacta proportio lentis & cavi ut integram possiderem Artem, quam quomodo nactus fuerim ne graveris intelligere seriem fati, quae tibi profutura erat. Anno 1611. cum Germaniam iter suscepissem, atque Oenipontum vulgo Ispruch, pervenissem ubi Serenissimus Dominus Maximilianus Archidux Austriae residet, diversorium exeunti occurrit illius domesticus, qui percontatus unde, & quo tenderem, atque ego declarassem animi sententiam, post salutationes & aliae humanitatis officia, secessit: sub vespera dum mensae accumberem adfuit servus, cum scheda monens ne praeterirem insalutato Principe, & condicta hora, mane perduxit ad illum, ad cuius conspectum cum ex more procidissem, & manum exosculatus post nonnulla praefatus, praebuit ansam prolixoris sermonis, post alia incidit sermo de Telescopia Galilaei, de quo cum vera narrassem, & verba facerem de proportione, duxit in Porticum ubi erat Mensa, in qua descripta & delineata erat proportio Galilaei, quam ipsemet miserat ad Maximilianum Bavarum Coloniae Archiepiscopum & Electorem: ipse autem Elector per Ioannem Zuchmesserum Mathematicum ac suum cubicularium ad Serenissimum Archiducem tansmiserat [*sic*]. Hanc cum attente considerassem, conversus ad Principem dixi eandem me prorsus habere, sed ex Hispania attulisse atque productis cartaceis mensuris peripheriam, primae differentiae convexi quae apud Rogetos est forma primi spicilli (vulgo *di vista commun*)[31] superposui, & erat eiusdem circuli sectio. Oculis etiam dimensus eram peripheriam globuli ad cavum spicillum, nec expectans Principis verba, produxi quoque peripheriam cartaceam quae est septima, & postrema cavi spicilli apud eosdem Rogetos, & imposita, omnino eadem reperta est.

[Here follows Sirtori's figure representing the radius of curvature of Galileo's lens.]

Ut Princeps vidit me easdem habere proportiones quas Elector Coloniensis miserat, interrogavit qua ratione posset hoc instrumentum perfici; Possem ego

[30] It was at this dinner that the word *telescopium* was first used. See Edward Rosen, *The Naming of the Telescope* (New York, Abelard-Schuman, 1947).

[31] See p. 11, above.

Serenissime Princeps, respondi, modo haberem ferramenta iuxta has proportiones: dati sunt Artifices, & interpretes cum mandatis, ut praecipienti praesto essent: Inter multos in ea regione vix unus repertus est qui intelligeret quid, aut quomodo esset faciendum. Interim saeviens pestis monebat mutare solum. Itaque salutato Principe prono flumine Viennam devehor, ubi totum me dedi huic studio, ut ferramenta perficerem, nulli labori, aut sumptui parcens. Scio Principes & potentes viros assequi non potuisse quod ego fato, sive sorte quadam sum assequutus, in Hispania Artem, & proportionem eandem circa quam laborarunt, & adhuc laborant Mathematici: Oeniponti eandem proportionem tanti Principis authoritate confirmatam. Nunc vero qua ratione & arte ferramenta paraverim, dicam ut possint & similia habere qui voluerint.

In the year 1609 [*sic*] there appeared a genius or some other man, as yet unknown, of the race of Hollanders, who, in Middelburg in Zeeland, visited Johannes Lippersein, a man distinguished from others by his remarkable appearance, and a spectacle-maker. There is no other [spectacle-maker] in that city, and he ordered many lenses to be made, concave as well as convex. On the agreed day he returned, eager for the finished work, and as soon as he had them before him, raising two of them up, namely a concave and a convex one, he put the one and the other before his eye and slowly moved them to and fro, either to test the gathering point or the workmanship, and after that he left, having paid the maker. The artisan, by no means devoid of ingenuity, and curious about the novelty, began to do the same and to imitate the customer, and quickly his wit suggested that these lenses should be joined together in a tube. And as soon as he had completed one, he rushed to the court of Prince Maurice and showed him the invention. The prince had one [*or,* had been acquainted with one] before, and lest it should be suspected that [the device] was of military value, and very necessary, had kept it a secret. But now that he found by chance that it had become known he disguised [his prior knowledge], rewarding the industry and good intentions of the artisan. Thence the novelty of so great a thing was spread through the whole world, and many other telescopes were made. But none of those turned out better or more apt than the first one (which I have seen and handled), so that you may say that not only the Arts, but Nature herself, brings about all things so that they may serve great princes. It was also related that this invention was nothing but two lenses put in a tube. And since Porta had made mention of this business in his *Magia naturalis,* although obscurely, and had spoken of it to many, in my presence, it appeared that this conception was in the minds of many men, so that once they heard about it, any ingenious person began trying to make one, without [the help of] a model. Others, Dutchmen, Frenchmen, Italians from everywhere rushed forward driven by the desire for gain, and there was no one who would not make himself the inventor. In the month of May a Frenchman rushed into Milan and offered such a telescope to the Count of Fuentes. He said that he was an associate of the inventor from Holland. When the count had given it to a silversmith to put it in a silver tube, it came into my hands. I handled it and examined it, and made similar ones, in which I observed that many inconveniences occurred because of the glass. I therefore went to Venice in order to obtain a supply [of lenses] from the artisans there and, being still unskilled in the art, I delivered a finished lens to someone so that he could make similar ones. I squandered some money uselessly and lost the lens, having learned nothing more than that the business is to be perfected by chance and by the laborious selection of lenses. As it happened, when I acquired one, I imprudently ascended the tower of Saint Mark, in order to try it out at a distance. Someone, having descried the novelty from the square, pointed it out to others, and a throng of noble youths, filled with curiosity, rushed upstairs and almost overwhelmed me—although they were well-behaved—and having courteously asked for the telescope, they began to look through it, passing it from one to the other. Tired by this delay of almost two hours, and by the unexpected occurrence, at length each one was summoned home by his empty stomach, and so the crowd began to grow thinner and I began to breathe again. The next day, remembering the perils of the previous day, and fearing that the same would happen if they were to find out my lodgings, concerning which they were inquiring as they were departing, I left. Yet, I was annoyed by this one thing, that I still was uncertain about the art, and had to acquire it with so much effort, and I thought of how I might master it. Meanwhile the rumor increased, and many lies went around (which whetted the appetite) about how it was asserted in the letters of merchants that in the Netherlands and Spain telescopes of this sort were to be found by which a man could be recognized at three miles. I took a trip to Spain, in the belief that all exceptional things would be obtainable there more surely and more quickly. When I came to Gerona, someone found out that I had the kind of glass that everyone was talking about. Soon, a certain curious architect appeared, asking whether he could see my telescope. Repelled by the man's aggressiveness, I began to refuse; he began to urge me anew, and would not go away, so that I came to suspect that the man was actually dedicated to the art. Nor was I wrong, for when he had looked at a remote tree as long as he wished, he further asked me to permit him to examine, take out, and handle the glasses. I agreed, knowing that because of his age he was unequal to the task, should he seek to imitate it. After he had handled and diligently considered the glasses, he led me to his lodgings and, having locked the door, he revealed the tools of the art, completely eaten away by rust. At one time he had been a spectacle-maker, and the entire art was buried there. As I felt myself won over by that favor of the expert in the art, I gave myself entirely into his friendship, and poured out my secret to him quite freely. He, moreover, showed me the outlines of the art, described in a book, and upon my asking, permitted me to copy out the proportions by means of only three points. After that it was not difficult for me to figure out the entire [series], and then, having examined the matter diligently, improved it by labor in daily experiments and expenses, and confirmed it, to perfect it and reduce it into this table, which I disclose to you. Our architect, as I learned later, was a brother of Roget of Burgundy, who at one time lived near Barcelona, a man of great diligence, who first introduced and established the art in Spain. He begot three sons. One of them, dedicated to letters and religion, devoted himself to the order of St. Dominic, and this monk wrote down the art. Nowhere was this art more exact than among those Roget brothers. Already I seemed to have learned the art whose outlines only I possessed. But things had turned out so nicely according to my wishes, that I hoped shortly to be able to perfect a telescope, and so, hastening to the royal court on private business, and quickly freed from there, I returned to my homeland, where I saw to the construction of some tools, and having hired some workers, began to observe how they practiced their art and to train my hand little by little. After long work and effort, when I made a trial of the lenses, it was especially evident to me that, from the same form, almost all lenses had turned out unequal to each other. When I diligently examined the cause of this, I found out that it was mainly due to the inequality of the form, but also to the varied thickness of the glass, and lastly also to the unsteadiness of the hand. With my hopes and efforts still frustrated, I went to Rome, where I had learned an abundance of all arts flourished and, moreover, many distinguished and celebrated men were dedicated to this study. Nothing more fortunate could have happened to me. For Galileo was there with his unforgettable telescope. By chance, on a certain day, Prince Federigo Cesi, Marquis of Monticello, a learned man and benefactor of the sciences, had invited him to dinner in the vineyard which is called Malvasia, and besides him some other learned men.[30] Before sunset, when they arrived there, they began to look through the telescope at the inscription of Pope Sixtus V above the Lateran portal, which was about a mile distant. I took my turn and looked and read the inscription to my satisfaction. Later that night, after dinner, we observed Jupiter and the motion of his companion stars, after which, sufficiently invigorated by the sight of such brilliance and by the curiosity of the matter, they withdrew in order to examine the telescope. And Galileo himself, in order to satisfy their curiosity, took out the lens and the concave glass, and showed them openly. In the meantime, I examined and measured the tube, and then also handled and inspected the lens, so that I can with confidence, from skilled experience, report what kind of thing it is. The exact proportion of the lens and the concave [glass] was the only thing to prevent me from possessing the whole art. You may wish to learn the series of events by which I became master of this, since it may be useful to you. In the year 1611, when I had made a journey to Germany, and had come to Innsbruck, where His Serene Highness Maximilian, Archduke of Austria, resides, leaving my lodgings, I chanced to meet a servant of his who inquired where I was from and where I was going, and I revealed my intentions to him. After greetings and other gestures of courtesy, he left. In the evening when I sat down at the dinner table, a servant appeared with a note advising me not to neglect to greet the prince, and at the appointed hour in the morning he led me to him, at whose sight I abased myself in the usual way, and kissed his hand. After some introductory remarks, he presented an opportunity for a longer conversation. After other [topics], there was a discussion about Galileo's telescope concerning which I told him the truth, and I made mention of its design. He took me to a porch where there was a board on which Galileo's design was described and drawn, the design which Galileo himself had sent to Maximilian of Bavaria, the Archbishop and Elector of Cologne. And the Elector himself had sent it to His Serene Highness the Archduke by means of Johannes Zuckmesser, a mathematician and his chamberlain. When I had studied it attentively, I turned to the prince and said that I possessed exactly the same thing, but had brought it from Spain. And having produced my paper patterns, I superimposed the outline of the first difference of the convex [lens], which according to the Rogets is the shape of the first lens (in Italian *of common vision* [i.e., a convex lens used for the correction of presbyopia]),[31] and it was a section of the same circle. With my eye I had also measured the outline of the little globe at the concave lens, and not waiting for the word of the prince, I also produced the paper pattern which is the seventh and last of the concave lens according to the same Rogets, and laid it on [the board], and it was found to be exactly the same.

As the prince saw that I had the same proportions which the Elector of Cologne had sent to him, he asked how this instrument could be made. "I could do it, Most Serene Prince," I replied, "provided that I had iron forms according to these proportions." Craftsmen and interpreters were provided, with orders that they be ready to be instructed. Among the many in that region, hardly one was to be found who understood what was to be made, or how. In the meantime, a violent plague made it advisable for me to move on, and so, having said goodbye to the prince, I was carried downstream to Vienna, where I gave myself wholly to the purpose of making the forms, sparing no labor or expense. I know that princes and great men could not attain what I, by some chance or fate, attained in Spain: that same skill and design which mathematicians have worked on and still strive for; in Innsbruck, the same design confirmed by the authority of so great a prince. Now I shall relate the method and the art by which I prepared the iron tools, so that those who may want to have similar ones may be able [to make them].

Galileo, *Il saggiatore* (1623). Text taken from *Le opere di Galileo Galilei* **6**: pp. 257–259.

Segue appresso, e, non so quanto opportunamente, s' induce a chiamare il telescopio mio allievo, ma a scoprire insieme come non è altrimenti mio figliuolo. Che fate, Sig. Sarsi? Mentre voi sete su 'l maneggio d' interessarmi in oblighi grandi per li beneficii fatti a questo ch' io reputavo mio figliuolo, mi venite dicendo che non è altro ch' un allievo? Che rettorica è la vostra? Avrei più tosto creduto che in tale occasione voi aveste avuto a cercar di farmelo creder figliuolo, quando ben voi foste stato sicuro che non fusse. Qual parte io abbia nel ritrovamento di questo strumento, e s' io lo possa ragionevolmente nominar mio parto, l' ho gran tempo fa manifestato nel mio Avviso Sidereo, scrivendo come in Vinezia, dove allora mi ritrovavo, giunsero nuove che al Sig. Conte Maurizio era stato presentato da un Olandese un occhiale, col quale le cose lontane si vedevano così perfettamente come se fussero state molto vicine; nè più fu aggiunto. Su questa relazione io tornai a Padova, dove allora stanziavo, e mi posi a pensar sopra tal problema, e la prima notte dopo il mio ritorno lo ritrovai, ed il giorno seguente fabbricai lo strumento, e ne diedi conto a Vinezia a i medesimi amici co' quali il giorno precedente ero stato ragionamento sopra questa materia. M' applicai poi subito a fabbricarne un altro più perfetto, il quale sei giorni dopo condussi a Vinezia, dove con gran meraviglia fu veduto quasi da tutti i principali gentiluomini di quella republica, ma con mia grandissima fatica, per più d' un mese continuo. Finalmente, per consiglio d' alcun mio affezzionato padrone, lo presentai al Principe in pieno Collegio, dal quale quanto ei fusse stimato e ricevuto con ammirazione, testificano le lettere ducali, che ancora sono appresso di me, contenenti la magnificenza di quel Serenissimo Principe in ricondurmi, per ricompensa della presentata invenzione, e confermarmi in vita nella mia lettura nello Studio di Padova, con dupplicato stipendio di quello che avavo per addietro, ch' era poi più che triplicato di quello di qualsivoglia altro mio antecessore. Questi atti, Sig. Sarsi, non son seguiti in un bosco o in un diserto: son seguiti in Vinezia, dove se voi allora foste stato, non m' avreste spacciato così per semplice balio: ma vive ancora, per la Dio grazia, la maggior parte di quei Signori, benissimo consapevoli del tutto, da' quali potrete esser meglio informato.

Ma forse alcuno mi potrebbe dire, che di non piccolo aiuto è al ritrovamento e risoluzion d' alcun problema l' esser prima in qualche modo reso consapevole della verità della conclusione, e sicuro di non cercar l' impossibile, e che perciò l' avviso e la certezza che l' occhiale era di già stato fatto mi fusse d' aiuto tale, che per avventura senza quello non l' avrei ritrovato. A questo io rispondo distinguendo, e dico che l' aiuto recatomi dall' avviso svegliò la volontà ad applicarvi il pensiero, che senza quello può esser ch' io mai non v' avessi pensato; ma che, oltre a questo, tale avviso possa agevolar l' invenzione, io non lo credo: e dico di più, che il ritrovar la risoluzion d'un problema segnato e nominato, è opera di maggiore ingegno assai che 'l ritrovarne uno non pensato nè nominato, perchè in questo può aver grandissima parte il caso, ma quello è tutto opera del discorso. E già noi siamo certi che l' Olandese, primo inventor del telescopio, era un semplice maestro d'occhiali ordinari, il quale casualmente, maneggiando vetri di più sorti, si abbattè a guardare nell' istesso tempo per due, l' uno convesso e l' altro concavo, posti in diverse lontananze dall' occhio, ed in questo modo vide ed osservò l' effetto che ne seguiva, e ritrovò lo strumento: ma io, mosso dall' avviso detto, ritrovai il medesimo per via di discorso; e perchè il discorso fu anco assai facile, io lo voglio manifestare a V.S. Illustrissima, acciò, raccontandolo dove ne cadesse il proposito, ella possa render, colla sua facilità, più creduli quelli che, col Sarsi, volessero diminuirmi quella lode, qualunqu' ella si sia, che mi si perviene.

Fu dunque tale il mio discorso. Questo artificio o costa d' un vetro solo, o di più d' uno. D' un solo non può essere, perchè la sua figura o è convessa, cioè più grossa nel mezo che verso gli estremi, o è concava, cioè più sottile nel mezo, o è compresa tra superficie parallele: ma questa non altera punto gli oggetti visibili col crescergli o diminuirgli; la concava gli diminuisce, e la convessa gli accresce bene, ma gli mostra assai indistinti ed abbagliati; adunque un vetro solo non basta per produr l' effetto. Passando poi a due, e sapendo che 'l vetro di superficie parallele non altera niente, come si è detto, conclusi che l'effetto non poteva nè anco seguir dall' accoppiamento di questo con alcuno degli altri due. Onde mi ristrinsi a volere esperimentare quello che facesse la composizion degli altri due, cioè del convesso e del concavo, e vidi come questa mi dava l' intento: e tale fu il progresso del mio ritrovamento, nel quale di niuno aiuto mi fu la concepita opinione della verità della conclusione.

Taken from Stillman Drake and C. D. O'Malley, *The Controversy on the Comets of 1618* (Philadelphia, University of Pennsylvania Press, 1960), pp. 211–213.

Immediately after this, though perhaps not very opportunely, he is induced to call the telescope my foster child and to reveal at the same time that it is not my offspring in any other way. Now how is this, Sig. Sarsi? When you are busy placing me under great obligations for benefits showered upon this supposed child of mine, must you go on to tell me that it is only a foster child? Is this rhetorically sound? I should have thought rather that on such an occasion you would have tried to make me believe it my own child even if you had been certain that it was not. The part which I played in the discovery of this instrument, and whether I may reasonably lay claim to it as my offspring, I set forth long ago in my *Starry Messenger.* I wrote that in Venice, where I happened to be at the time, news came that a Hollander had presented to Count Maurice [of Nassau] a glass by means of which distant things might be seen as perfectly as if they were quite close. That was all. Upon hearing this news, I returned to Padua, where I then resided, and set myself to thinking about the problem. The first night after my return, I solved it, and the following day I constructed the instrument and sent word of this to the same friends in Venice with whom I had been discussing the subject the previous day. Immediately afterwards, I applied myself to the construction of another and better one, which I took to Venice six days later; there it was seen with great admiration by nearly all the principal gentlemen of that republic for more than a month on end, to my considerable fatigue. Finally, at the suggestion of one of my friendly patrons, I presented it to the ruler in a full meeting of the Council. How greatly it was esteemed by him, and with how much admiration it was received, is testified by ducal letters still in my possession which reveal the munificence of that serene ruler in recompense for the invention presented to him, reappointing and confirming me for life to my professorship at the University of Padua at double my former salary, which was then more than triple that of some of my predecessors. These acts did not take place in some forest or desert, Sig. Sarsi; they happened in Venice, and if you had been there you would not dismiss me thus as a simple schoolmaster. But by the grace of God, most of those gentlemen are still living there, and by them you may be better informed.

Perhaps someone will say, however, that in the discovery and solution of a problem it is of no little assistance first to be conscious in some way that the conclusion is true and to be certain that one is not attempting the impossible; and hence that my knowledge and certainty that a telescope had already been made were of so much help to me that without this I should perhaps not have made the discovery. To this I shall reply by making a distinction. I say that the aid afforded me by the news awoke in me the will to apply my mind to it, and without this I might never have thought about it; but beyond that I do not believe that such news could facilitate the invention. I say, moreover, that to discover the solution of a known and designated problem is a labor of much greater ingenuity than to solve a problem which has not been thought of and defined, for luck may play a large role in the latter while the former is entirely a work of reasoning. Indeed, we know that the Hollander who was first to invent the telescope was a simple maker of ordinary spectacles who in casually handling pieces of glass of various sorts happened to look through two at once, one convex and the other concave, and placed at different distances from the eye. In this way, he observed the resulting effect, and thus discovered the instrument. But I, incited by the news mentioned above, discovered the same by means of reasoning which, easy as it is, I wish to reveal to your Excellency. Recounting it thus where it is to the purpose, I may by its simplicity render less incredulous those people who, like Sarsi, may wish to diminish whatever praise there is in it that belongs to me.

My reasoning was this. The device needs either a single glass or more than one. It cannot consist of one alone, because the shape of that one would have to be a convex (that is, thicker in the middle than at the edges), or concave (that is, thinner in the middle), or contained between parallel surfaces. But the last named does not alter visible objects in any way, either by enlarging or reducing them; the concave diminishes them; and the convex, while it does indeed increase them, shows them very indistinctly and confusedly. Therefore, a single glass is not sufficient to produce the effect. Passing next to two, and knowing as before that a glass with parallel faces alters nothing, I concluded that the effect would still not be achieved by combining such a one with either of the other two. Hence I was restricted to trying to discover what would be done by a combination of the convex and the concave, and you see how this gave me what I sought. Such were the steps in my discovery, in which I was not at all assisted by the conception that the conclusion was true.

Journal tenu par Isaac Beeckman de 1604 à 1634 publié avec une introduction et des notes par C. De Waard (4 v., The Hague, Martinus Nijhoff, 1939–1953) **3**: p. 376, June, 1634.

Johannes Sacharias seght dat syn vader [32] den eersten verrekycker maeckte hier te lande an° 1604 naer eene van eenen Italiaen daerop stont an° 190.[33]

Johannes Sachariassen says that his father made the first telescope in this country in the year 1604, after one belonging to an Italian which bore the date *anno* 190.[33]

René Descartes, *La dioptrique* (1637). Text taken from *Œuvres de Descartes publiées par Charles Adam & Paul Tannery* (Paris, 1897–1913) **6**: pp. 81–82.

Mais, a la honte de nos sciences, cete invention, si utile & si admirable, n'a premierement esté trouvée que par l'experience & la fortune. Il y a environ trente ans, qu'un nommé Iaques Metius, de la ville d'Alcmar en Hollande, homme qui n'avoit iamais estudié, bien qu'il eust un pere & un frere qui ont fait profession des mathematiques, mais qui prenoit particulierement plaisir a faire des miroirs & verres bruslans, en composant mesme l'hyver avec de la glace, ainsi que l'experience a monstré qu'on en peut faire, ayant a cete occasion plusieurs verres de diverses formes, s'avisa par bonheur de regarder au travers de deus, dont l'un estoit un peu plus espais au milieu qu'aus extremités, & l'autre au contraire beaucoup plus espais aus extremités qu'au milieu, & il les appliqua si heureusement aus deus bouts d'un tuyau, que la premiere des lunettes dont nous parlons, en fut composée. Et c'est seulement sur ce patron, que toutes les autres qu'on a veües depuis ont esté faites, sans que personne encore, que ie sçache, ait suffisanment determiné les figures que ces verres doivent avoir.

To the disgrace of our sciences, that invention, so useful and admirable, was at first found only through experience and luck. It was about thirty years ago that a man named Jacob Metius, of the city of Alkmaar in Holland, a man who had never studied, although he had a father and a brother who made a profession of mathematics, but who took particular pleasure in making burning mirrors and glasses, even making them out of ice in the winter, as experience has shown they can be made, having on that occasion several glasses of different shapes, decided through luck to look through two of them, of which one was a little narrower in the middle than at the edges, and the other, on the contrary, much thinner at the edges than in the middle. And he put them so fortunately in two ends of a tube, that the first of the telescopes, of which we are speaking, was put together. And it is entirely based on this model that all the others which have been seen since have been made without anyone yet, as far as I know, having sufficiently determined the shapes that those glasses ought to have.

Antonius Maria Schyrlaeus de Rheita, *Oculus Enoch et Eliae seu radius sidereo-mysticus* (Antwerp, 1645), part I: pp. 337–338.

Enim vero Batavus quidam Anno 1609. conditione quidem humilis, arte autem perspicilliarius, nomine Ioannes Lippensum Zelandus, casu concavum cum convexo specillo coniungens, sicque utrumque oculo applicans, haud sine admiratione animadvertit obiecta magnopere augeri, oculoque mirum in modum propinqua reddi. Quo viso, tum utrumque vitrum in tubum seu canalem debita proportione disponit, & ita forte fortuito transeuntibus turris gallum quasi ex ioco, spectandum praebet. Increbuit huius novitatis paulatim fama, alliciuntur catervatim viatores, novum & inauditum spectaculum haud sine curiositate & stupore spectantes: casus itaque in utilitatem transit, iocus in rem seriam vertitur. Enimvero Marchio Spinola tunc

[32] I.e., Sacharias Janssen.

[33] De Waard argued that Beeckman simply forgot the *5* (*De uitvinding der verrekijkers,* p. 155). Stillman Drake has suggested that the *1* may be a degenerate *5,* and that the number could then be read as *590* (personal correspondence). The word *anno* makes it clear that a date was meant, and *1590* surely is the one meant.

temporis Hagae Comitis suspensionem armorum cum statibus tractaturus degens novum instrumentum illud visum pecunia sibi comparatum, Serenissimo Archiduci Alberto piae memoriae offert. Interim res defertur ad proceres, vocatur homo ille perspicilliarius, tubumque alterum a se factum, sat magno pretio, ea tamen inhibitione & onere vendere cogitur, ne deinceps simile instrumentum elaboret vendatque &c. Quomodo & hoc nobilissimum casuale inventum instrumentumque, in abscondito repositum celatumque remansisset nisi divina ita disponente voluntate, iam aliunde, uti diximus prius fuisset transmissum, & in Brabantina aula divulgatam.

Coepit itaque paulatim ulterius in Italiam, Germaniam, & Galliam divulgari, magisque perfici, donec ad eam, in qua nunc temporis est, perfectionem tandem deductum fuerit. Et profecto quemadmodum, embrio prius informe quid & imperfectum est, tum vero paulatim perficitur, donec in hominem excrescat perfectum, ita hoc specillum primo fuit quasi informe & imperfectum, tum a Galilaeo, aliisque ad maiorem perfectionem deductum est: Tandem autem ut spero omnibus nostro tempore numeris absolvendum.

In the year 1609, a certain Dutchman of humble origin and, in fact, a spectacle-maker by profession, named Ioannes Lippensum of Zeeland, combining by chance a concave with a convex lens, and applying them to his eye, noticed not without surprise that objects were greatly enlarged, and that they were brought astonishingly close to the eye. Having seen this, he then put both glasses in a tube or pipe of the required size, and so he offered those who happened to pass by, as a joke, a look at the weather cock on the tower. The rumor of this novelty spread gradually, and travelers were attracted in droves to see the new and unheard of spectacle, not without curiosity and amazement. The accident was thus transformed into usefulness, and a joke was turned into a serious thing. In fact, the Marquis Spinola, visiting The Hague at that time in order to negotiate the suspension of hostilities with the States [of the Netherlands], having seen the new instrument, bought one for himself and presented it to the Most Serene Archduke Albert of pious memory. In the meantime, the matter was reported to the rulers, and that man, the spectacle-maker, was summoned, and another tube was made by him at a rather high price, with the condition and obligation, however, that he should not, in the future, make and sell a similar instrument, etc. In this way, the very noble instrument, which had been invented fortuitously, would have remained locked away as a secret if it had not already, because God's will so disposed matters, been sent elsewhere, as we said already, and made known at the Court of Brabant.

And so it gradually became widely known abroad, in Italy, Germany, and France, and was further perfected until it was at length brought to that degree of perfection which it has attained at the present time. And indeed, just as an embryo is first something shapeless and imperfect, and is then gradually perfected until it grows into a perfect man, so this instrument was at first, as it were, shapeless and imperfect, until it was brought to a greater perfection by Galileo and others. And finally, it will be completed in every respect in our own time, as I hope.

Willem Boreel to the City Council of Middelburg, 8 January, 1655. Original destroyed. Text taken from de Waard, *Uitvinding,* pp. 10–11.

Edele, erntfeste, wyse, hoochgeleerde,
seer voorsienige Heeren.

Gelijc ick alletijt seer geern sal toebrengen alles wat can strecken tot eere ende renommée van UweEd. Erntf. stadt ende mijn vaderlandt, soo compt my nu voor eenige goede occasie daertoe. UweEd. Erntf. is ten vollen wel bekent de heerlijcke nieuwe inventie van de verdesiende brillen ofte verdekykers, door middel van dewelcke de mathematiques grootlijcx sijn geholpen geworden, maer oock vele schoone heerlyke schepsels Gods, soo in hemel als op aerde, sijn ontdeckt en bekendt geworden, die noeit menschen ooghe voor desen, dat men weet, en heeft gesien. Ydereen souckt aen hem te trecken d'eere van die vond, Galileus de Galileis, Velserus,[34] Metius, van Alcmare deer haer daervan wel aennemen en toeschryven, insonderheit den laetsten, hoewel sy alle, mijns erachtens, vermeerderaers en illustrateurs en sijn van de gevonden sake binnen UweEd. Erntf. stadt. Soo ick wel geïnformeert ben en wel onthouden can, soo hebbe ick de man gekendt, gesien en gesproken in myne joncheyt, die gesegt werde d eerste inventeur geweest te sijn, hoewel met wat imperfectie, van de voorschreven verrekykers, die hy naemaels van tijt tot tijt seer heeft verbeterd; als oock die geleerde en andere ervaren mannen, die deselve als voorschreven hebben geïllustreert. Dese man woonde tot Middelburgh in de Capoenstraete comende van de Groenmart aen de slynckerhandt, ontrent het midden van de straete, in de huyskens, die tegens de Nieuwe kercke aenstaen; was een man van geringhe middelen, hadde eene sobere wynckel, veel kinders, die ick daerna noch hebbe gesien, ouder wordende ende tot Middelburch comende. Indien Uwe Ed. Erntf. myne memorie by wetlijck ondersouck bevinden goet te wesen, ende dat d'eere van die inventie de stad Middelburgh toecome, soo versoucke, dat ick daervan by toegesondene documenten versekert moge werden; ick hebbe hier nu goede occasie om d'eere van UweEd. Erntf. stadt te connen verbreyde, en na dienstighe gebiedenisse ben ende sal altijt blyven,

Mijn Heeren,
UweEd. Erntf.
dienstwillighe ingeboren
W. Boreel.

Tot Paris desen 8 January
1655.

Noble, Honorable, Wise, Learned,
Very Providential Gentlemen.

As I am always eager to contribute anything that can add to the honor and renown of your city and my father-

[34] Welser never claimed he invented the telescope, nor did anyone ever ascribe the invention to him.

land, a good opportunity now presents itself to me. You are, I suppose, fully familiar with the wonderful new invention of far-seeing spectacles or telescopes, by which the sciences have been greatly helped, but by means of which also many beautiful, wonderful creations of God, in the heavens as well as on Earth, have been discovered and become known, which before this had never been seen by human eyes, as far as we know. Everyone seeks to claim the honor of that invention for himself. Galileo Galilei, Welser,[34] and Metius of Alkmaar have assumed the honor or it has been ascribed to them, especially the last. But in my opinion they are all popularizers and illustrators of the thing invented in your city. According to what I have been told and what I can remember, in my youth I knew, saw, and spoke to the man who is said to have been the first inventor of the said telescopes—although with some imperfections—which afterwards, from time to time, he improved greatly, as indeed was done by those learned men and other skilled men, who have embellished the same. This man lived in Middelburg in the Capoen Street, on the left side coming from the Vegetable Market, in about the middle of the block, in the little houses against the new church. He was a man of small means, had a modest shop, and many children, whom I still saw afterwards when I came back to Middelburg when I was older. If after legal investigation you find that my recollections are correct, and that the honor of the invention belongs to the city of Middelburg, I request that I be notified of this by means of documents. At the present time I have a good opportunity here to further the honor of your city, and after offering you my respects, I am, and shall always remain, Gentlemen,

Your Honors'
Obedient native of Middelburg,
W. Boreel.

At Paris, this 8th of January
1655.

Johannes Sachariassen to the City Council of Middelburg, 30 January, 1655. Original destroyed. Text taken from de Waard, *Uitvinding,* p. 140.

Anno 1590 is de eerste buyse gemaeckt en geïnventeert binnen Middelburgh in Zeelant van Zacharias Jansen, ende de langste waerr doen ter tijt 15 à 16 duym, waervan datter 2 wech vereert werden: de eene aen den prins Mourytsyus en de ander aen hertogh Albertus. —De destancy van 15 à 16 duym is soo lange gegebruyckt [*sic*] geweest tot het jaer 1618; doen hebbe ick met mijn vader, hierboven vernoumpt, de lange buysen geïnventert, die men gebruyckt om by nachte te sien in de sterren en de maenne,[35] daer veel in te spekeleren is. Anno 1620 heeft Meetsyus [36] een van onse buysen bekommen, dewelcke hy naergekonterfeyt heeft, voor sooveel als hij gekonnen heeft: desgelickx heeft oock Cornelis Drybbel gedaen; als wy dese instermenten practyseerden, woonden wy op het kerckhof, daer nu de venduysy is. Waerre Reynnier Ducartes en Cornelis Dribbel en Johannes Loof int leven, die souden getuygen daervan konnen wesen, dat ick de eerste lange buysen hebbe geïnventert; vorder en kan ick mijn Heeren geen naeder onderricht daervan doen.

In Middelburgh den 30 Jannewary 1655.
UE. W. onderdaene dienaer
Johannis Sachariassen.

In the year 1590 the first tube was made and invented in Middelburg in Zeeland by Sacharias Janssen, and at that time the longest were 15 to 16 inches [long]. Of these, two were presented, the one to Prince Maurice and the other to [Arch]duke Albert. — The length of 15 to 16 inches was in use until the year 1618; then I and my father, named above, invented the long tubes which are used at night for seeing the stars and the Moon,[35] about which there is much speculation. In the year 1620, Metius [36] obtained one of our tubes, which he copied as well as he could; Cornelis Drebbel did the same. When we were making these instruments we lived in the churchyard, where the market is now located. If René Descartes and Cornelis Drebbel and Johannes Loof were still alive, they could testify that I invented the first long tubes. I cannot give you Gentlemen further information.

At Middelburg, the 30th of January, 1655
Your Honors' Obedient Servant
Johannes Sachariassen

The City Council of Middelburg to Willem Boreel, 3 March, 1655. Original destroyed. Text taken from de Waard, *Uitvinding,* pp. 16–17, 11–13, 14–15.

Edele, gestrenge, eernfeste, wyse,
seer voorsienige Heer.

De missive van UweEdelheyt, geschreven den 8[en] January, is ons wel ter hant gekomen, ende bedancken deselve gansch hooghelijcke voor de toegenegene sorge om d'eere van dese stadt te laeten verbreyden door de heerlijcke inventie van de verrekijckers off verresiende brillen, waervan den eersten vinder alhyer soude hebben gewoont. Wy hebben geern omme UEdelheyts goede intentie op te volgen naersticheyt laeten aenwenden om den man uyt te vinden, ende is ons ten eersten daertoe aenleydinge gegeven door onsen medebroeder in wetthe Jacob Blondel, oud LXV jaeren, in effecte confirmerende hetgeene UEdelheyt diesaengaende was schryvende; edoch naer verder en curieus ondersouck hebben tot nochtoe niet anders konnen bekomen, als hetgeene Uwe Edelheyt uyt het inleggende sal gelieven te sien; [37] ende gelijck als diverse haer geerne d'eere van dese treffelijcke inventie souden aenmaetigen, soo dunckt ons mede, dat alhier eenen Johannes Sacharias insgelijcx wilt doen,[38] dan op het bericht van den voornoemden heer onsen confrater Blondel, accoorderende met Uwe Edelheyt, misgaders de verklaeringen van Jacob Willemsen ende Eewoud Kien dunckt ons, dat hy sigh abuseert en qualijck moet hebben onthouden; edoch in allen gevalle oordeelen sijn bericht mede astruërende alsdat d'inventie uyt dese

[35] See above, p. 23, note 26.

[36] In all likelihood this was *Adriaen* Metius, the professor at Franeker.

[37] See the two following letters.

[38] See preceding letter and pp. 57–58, below.

onse stadt is voortgekomen, waervan niet meerder voor als noch; sullen soo yets naerders konnen bekomen niet mancqueren dat over te senden, en onderentusschen aen Uwe Edelheyt laetende off en hoe d'inleggende gelieft te gebruycken, sullen seer geern verwachten andere occasiën om Uwe Edelheyt genegentheyt te mogen verschuldigen, waertoe Edele etc. den goeden Godt will verleenen Uwe Edelheyt langhduyrige gesontheyt en voorspoet.

Verblyvende
Uwer Edelheyts gunstige en
dienstwillige vrienden,
De borgemeesters, schepenen en raedt.

Den 3^{en} Meerte 1655.
t'Opschrift:
Aen den ambassadeur Boreel
tot
Parijs.

Noble, stern, honorable, wise,
very providential Sir.

Your letter, written on the eighth of January, has reached us safely, and we thank you very much for the trouble you are taking to further the honor of this city, through the wonderful invention of telescopes or far-seeing spectacles, of which the first inventor is alleged to have lived here. In order to follow up on your good intention, we have eagerly taken pains to find out about the man. And we were first informed by our colleague in the Law, Jacob Blondel, aged sixty-five years, who in effect confirmed what you wrote about this matter. But after further and diligent investigation, we have thus far been unable to obtain anything further than that which you will please find in the enclosed.[37] And just as diverse persons would like to appropriate the honor of this excellent invention, it appears to us also that a certain Johannes Sachariassen here would do the same.[38] For in view of the communication of the said gentleman, our colleague Blondel, and the declarations of Jacob Willemsen and Eeuwoud Kien, we think that he [i.e., Johannes Sachariassen] is mistaken and must have remembered wrongly. But in any case, we think that his communication supports the contention that the invention originated in this city. Should we be able to obtain something further, we shall not fail to send it to you, leaving it up to you, in the meantime, how you wish to use the enclosed. We shall eagerly await other opportunities to incur the debt of your affection, for which, Noble, etc., the good Lord will grant you longlasting health and prosperity.

Your well-wishing and
obliging friends,
The burgomasters, aldermen and Council.

The 3d of March, 1655
the Address
To the ambassador Boreel
at
Paris.

DEPOSITIONS IN FAVOR OF HANS LIPPERHEY APPENDED TO ABOVE LETTER

Burgemeesters, schepenen ende raden der stad Middelburgh in Zeeland, hebbende gedaen examineren en hooren de persoonen van Jacob Willemssen, concherge in de wisselbancq alhier, gaende in sijn 70 jaar, en Eeuwoud Kien, bode van dese stadt op Antwerpen, oudt zevenenzestigh jaren, mitsgaders Abraham de Jonge, mr. smit binnen deser stadt, oudt zevenenzeventigh jaren, over de kennisse, die sy gesamentlijck en yder van hun int bysonder hadden van seker persoon, die binnen deser stede soude gemaeckt hebben d'eerste verresiende brillen oft verrekijckers, dewelcke gevraeght zijnde, hebben getuycht ende verclaert als volght: En eerstelijck de voornoemde Jacob Willemsen seght den voorschreven persoon genaemt te sijn geweest Hans Laprey, ende dat deselve gewoont heeft in de Capoenstrate binnen deser stad, daer jegenwoordigh een stopper in woond, oft het naeste huys daeraen, sonder daerin te willen sijn begrepen, segt oock hem gekent te hebben, terwyle hy brillen maeckte, ende naderhand dat hy verrekijckers oft buyssen gemaeckt heeft, ende dat hetselve geleden is omtrend vijftigh jaren; dat denselven Laprey, na sijn memorie, omtrend de dartigh jaren is overleden en wel te weten denselven alhier binnen de stad Middelburgh is overleden, gevende voor redenen van welwetenschap, dat hy deponent vier à vijff huysen van het voornoemde huys heeft gewoont, en dat denselven Hans Laprey over d'eerste verrekijckers, die hy maeckte, een vereeringe heeft verkregen van den prins Maurits, soo hy dickmaels heeft hooren seggen. Ende den voornoemden Eeuwoud Kien deposerende verclaerde den naem van den persoon, die de brillen of verrekijckers plach te maken, geweest te sijn Hans Laprey van Wesel, en dat denselven binnen deser stadt gewoont heeft in de Cappoenstraete, tegens de Nieuwe kercke aen, in het huys daer de verrekijckers plegen uuyt te steken, naest het huys, daer nu het Serpent uuythanght, welcke huysen den voornoemde Laprey toegecomen hebben. Seght oock den gemelten Laprey omtrent het jaar 1610 de voornoemde verrekijckers heeft begonnen te maken en in den jare 1619 in de maend van October binnen deser stede overleden en begraven te sijn, voor redenen van wetenschap allegerende, dat hy den voornoemde Laprey's doghter in huywelijck heeft gehadt, dat oocq denselven Laprey soo aen de Staten als aen Sijn Excellentie prins Maurits van deselve verrekijckers vereert gehadt heeft ende eenige schenkagiën in recompense daerover gekregen te hebben, oock op sijn versoucq octroy geobtineert voor den tijdt van 3 jaaren.[39] Den voornoemde Abraham de Jonge, mede sijnne getuygenisse gevende, seght en verclaert den persoon, die deerste verrekijckers hier binnen deser stadt gemaecqt heeft, is genoemt geweest Hans, sonder synen toenaem perfectelijck onthouden te hebben, maer dat men hem int gemeen noemde Hans den brillemaker. Seght denselven gewoont te hebben in de Kapoenstraete binnen deser stede, sonder presycelijck te weten in wat

[39] Lipperhey was not granted a patent, see p. 42, above.

huys, ende dattet naer sijn onthoudt omtrent 5 a 46 jaeren is geleden, dat den voornoemde Hans deerste verrekykers maecte, dat hy deponendt denselven ettelijcke jaren te vooren, eer hy noch b[r]illen maeckte, gekent heeft, ende dat hy doen een metselaersknegh̄t was, gevende voor redenen van wetenschap, dat hy deposant in het huys, daer hy nogh woont, staende op de Wal alhier, wel 50 jaaren heeft gewoont ende als gebuyr aghter tlijck van den voornoemden Hans te begravenisse heeft medegegaen. Seght oock wel te weten ende meenichmael heeft hooren verhaelen, dat den voornoemde Hans voor Sijn Exsellentie Mauritius eenige buysen ofte verrekykers heeft gemaecqt. Des toirconde etc.

Simon van Beaumont.

The burgomasters, aldermen, and councillors of the city of Middelburg in Zeeland, ordered that the persons of Jacob Willemssen, porter at the exchange bank here, going into his 70th year, and Eeuwoud Kien, messenger of this city to Antwerp, aged sixty-seven years, as well as Abraham de Jonge, master smith in this city, aged seventy-seven years, be examined and heard about the knowledge which they had, collectively and individually, about a certain person who in this city allegedly made the first far-seeing spectacles or telescopes, and upon being asked they testified and declared as follows: And first, the said Jacob Willemssen says that the above-mentioned person was named Hans Laprey, and that the same lived in the Capoen Street in this city, where at present a darner lives, or in the next house, not wanting to be held to it. He also says that he knew him when he was making spectacles, and that afterwards he made tubes or telescopes, and that this is about fifty years ago; that the same Laprey died about thirty years ago, and that he knows that the same died here in the city of Middelburg, giving as his reason for knowing that he, the witness, lived four or five houses from the said house, and that the same Laprey received an honorarium from Prince Maurice for the first telescopes he made, as he has often heard it said. And the said Eeuwoud Kien, making his deposition, declared that the name of the person who used to make spectacles or telescopes was Hans Laprey of Wesel, and that the same lived in this city in the Capoen Street, next to the new church, in the house from which telescopes stick out, next to the house where now the Serpent hangs, which houses used to belong to the said Laprey. He also says that the said Laprey began making the said telescopes about the year 1610, and that in the year 1619, in the month of October, he died and was buried in this city, citing as reasons for knowing that he had been married to the said Laprey's daughter, and that also the same Laprey presented the same telescopes to the States[-General] as well as to His Excellency Prince Maurice, and received some recompenses in return, also getting his requested patent for the period of three years.[39] The said Abraham de Jonge, also witnessing, says and declares that the person who made the first telescopes here in this city was called Hans, not remembering his surname perfectly, but that he was usually called Hans the spectacle-maker. He says that the same lived in the Capoen Street in this city, without knowing exactly in which house, and that according to his recollections it was 45 or 46 years ago that the said Hans made the first telescopes; that he, the witness, knew him several years earlier, when he was not yet making spectacles, but was at that time a journeyman bricklayer, giving as reasons for knowing that he, the witness, has lived in the same house on the Wall [street] for fifty years, and as a neighbor walked behind the corpse in the funeral of the said Hans. He also says that he knows, and has heard it told many times, that the said Hans made some tubes or telescopes for His Excellency Maurice. In witness thereof, etc.

Simon van Beaumont

DEPOSITIONS IN FAVOR OF SACHARIAS JANSSEN APPENDED TO ABOVE LETTER

Burgemeesters, schepenen ende raden der stadt Middelburgh in Zeeland, hebbende gedaen hooren ende examineren de personen van Johannes Sachariassen, brillemaker binnen deser stede, oudt tweeënvijftigh jaeren, ende Sara Goedaerts, weduwe wylend Jacob Goedaerdt, oudt tweeenseventigh jaren, wonende int Goude kruys op de Kaye binnen deser stede, over de kennisse, die sylieden te samen ende yder int bysonder soude mogen hebben van seecker persoon, die binnen deser stede gemaeckt heeft d'eerste verresiende brillen oft verrekijckers, dewelcke gevraeght sijnde, getuyght en verklaert hebben als volght: Eerstelijck den voornoemde Johannes Sachariassen seght, die by sijnnen vader met namen Sacharias Janssen, soo hy meermaels heeft hooren seggen, in den jare XV^{c} tnegentigh alhier binnen Middelburgh sijn geïnventeert geweest, en dat de langhste buyse doen der tijt geweest is van vijfthien à sesthien duymen, datter twee van deselve gemaeckte verrekijckers vereert geweest sijn, d'eene aen den prince Mauritius ende d'ander aen den hertogh Albertus, welcke buysen van alsulcken lenghte sijn gebruyckt geweest tot den jare XVI^{c} aghthien incluys, als wanneer hy deposant verclaerde mettenselven synen vader de lange buysse geïnventeert te hebben, die men gebruyckt om by naghte te sien in de sterren ende manen; seght mede, dat eenen Meetsius[40] in den jare XVI^{c} twintigh, een van deselve verrekijckers bekomen hebbende, die nabootste, soo hy best conde; desgelijcx eenen Cornelis Dribbel. Voorts dat ten tyde van het practiseren van de voornoemde buysen sijnnen voornoemde vader gewoont heeft op het Kerckhof tegens de Nieuwe kercke, daer nu de vendue gehouden werdt.

Ende de voornoemde Sara Goedaerts, mede getuygenisse gevende, seght omtrendt twee- à vierenveertigh jaren geleden, sonder den precysen tijt te hebben onthouden, alhier buysen oft verrekijckers binnen deser stede gemaeckt wierden by Sacharias Janssen, haren broeder saliger, woonende by de Munte dight aen de Nieuwe kercke, gevende voor redenen van welwetenschap, dat sy den voornoemde haren broeder [met] deselve verrekijckers ontallijcke reysen heeft sien maecken.

Des t'oirconde hebben wy burghmeesters ende schepenen voornoemt desen gedaen segelen met den contrasegele der voorschreven stadt ende by een van

[40] See above, p. 55, note 36.

onse secretarissen laten onderteeckenen op den derden Martij XVI[c] vijfenvijftigh.

Simon van Beaumont.

The burgomasters, aldermen and councillors of the city of Middelburg in Zeeland ordered that the persons Johannes Sachariassen, spectacle-maker in this city, aged fifty-two years, and Sara Goedaert[s], widow of the late Jacob Goedaerdt, aged seventy-two years, living in the Golden Cross on the quay in this city, be heard and examined about the knowledge which they might have, together and individually, about a certain person who, in this city, made the first far-seeing spectacles or telescopes, and having been asked this, they testified and declared as follows: First, the said Johannes Sachariassen says that they were invented by his father Sacharias Janssen, here in Middelburg, in the year 1590, as he has often heard it told, and that at that time the longest tubes were fifteen to sixteen inches long, and that two of the said tubes were presented, one to Prince Maurice and the other to the [Arch]duke Albert, and that tubes of such lengths were used until the year 1618, when, as he, the witness, declared, he invented with the same father the long tubes which are used to see the stars and Moon at night. He also says that a certain Metius,[40] in the year 1620, having obtained one of the same telescopes, imitated it as well as he could, and likewise a certain Cornelis Drebbel. Further, that at that time of the making of the said tubes his father lived in the churchyard, against the new church, where now the market is held.

And the said Sara Goedaerts, also giving testimony, says that about forty-two or forty-four years ago, without recalling the time exactly, tubes or telescopes were made here within this city by Sacharias Janssen, her late brother, living near the Mint, close to the new church, giving as reason for her knowledge that she has seen her brother make numerous journeys with the same telescopes.

In witness thereto, we, the said burgomasters and aldermen, have sealed this with the counterseal of the said city, and let it be signed by one of our secretaries on the third of March, 1655.

Simon van Beaumont

Pierre Borel, *De vero telescopii inventore* (The Hague, 1655–1656), pp. 18–29.

Cap. VIII

Inventum Telescopii sibi omnes nationes arrogasse

Nulla fuit natio quae sibi Telescopii inventum admirandum non arrogarit: Galli enim, Hispani, Angli, Itali, Batavique rem suam facere contenderunt, ut patet ex *Sirturo* lib. de Telesc. cap 1. part. 1, [*sic*][41] cum tamen apud nullum eorum repertum, sed vere in Selandia Belgica, ut probabitur loco suo.

Quidam rem cognitam fuisse antiquis existimant, sed inter arcana custodita: at si alicui innotuerit, cognita Democrito fuit (magno illi Philosopho, cujus Philosophiam compono labore meo dignissimam) [42] qui viam lacteam, stellarum congeriem primus dixit, aliaque rara quae oculorum acie rimari arduum, imo impossibile erat, sed ejus inventum oblivio sepultum usque ad nostra tempora jacuit.

Alii *Bacconi* Anglo rem cognitam fuisse contendunt; Alii *Baptistae Portae,* qui quaedam de hac re obscure tamen dixisse videtur: nec desunt qui viro Sedanensi, *Crepii* vocato,[43] artifici eximio, hanc concedant: sed a nullo publici juris facta cum fuerint, jure de hoc dubitare possumus. Fuit quidam *Frater Paulus* Italus,[44] vir acerrimi ingenii, qui conspicilii fabricam agnovit, sed ex relatu illud habuerat, & omnia ex Selandia originem traxerunt, ut luce clarius demonstrabitur.

At major opinio pro *Drebellio,*[45] *Galilaeo* & *Metio* fuit, quibus omnibus hoc inventum a multis tribuitur, sibique ipsimet arrogare non erubescunt, cum tamen optime cunctis patescat, imo publicis testimoniis eos artificem Middelburgensem convenisse, vel id ab eo alio pacto mutuasse; maxima authorum graviorum pars Gallos, Italos, &c. hoc privant invento, & rem Batavis concedunt, sed adhuc hi non rem plane agnoverunt: accedunt quidem magis ad veritatem, quod circa hanc regionem rem inventam fuisse subolfecissent, sed maluerunt Mathematicis celeberrimis eam quam vili artifici tribuere.

Sic Cartesius in sua dioptrica inventorem Hollandum esse asseruit, & videtur Metium intelligere.

Nec defuerunt qui ex Hispania rem prodiisse ausi fuerunt [46] asserere, sed haec non minus obscura cum sint, ac de Viro Sedanensi,[47] cumque a paucissimis probata sint, ea non amplius inquiremus.

Cap. XI [*sic*]

Galilaeum non invenisse Telescopia, sed Selandos

Inter hanc inventorum conspiciliorum turbam primo insurgit Galilaeus, qui sibi inventum tribuit,[48] & pro vero Inventore huc usque inter multas nationes laudibus elatus fuit, seque ipsum propriis extulit encomiis, ut libello suo rogatorio ad Rempublicam Holland. oblato patet, unde Galilaei conspicilia vulgo vocato fuerunt.[49] Non tamen si quae ei debeatur gloria in inventi augmento ac perfectione, illa eum privandum esse censeo,

[41] *Telescopium: sive ars perficiendi,* see pp. 48, 50 above.

[42] At this time Borel was working on a *Vita ac philosophia Democriti,* which he never published. See Pierre Chabbert, "Pierre Borel (1620?–1671)," *Revue d'histoire des sciences et de leurs applications* 21 (1968): p. 342.

[43] See above, p. 43.

[44] Borel is probably referring to Fra Paolo Sarpi (1552–1623) who was perhaps the first person in Italy to learn about the invention of the telescope. See Stillman Drake, *Galileo studies,* p. 143.

[45] Borel is the first to mention Drebbel as a candidate.

[46] See pp. 49, 50, above.

[47] See above, p. 43.

[48] See above, p. 5, note 6.

[49] In 1636 Galileo submitted to the States-General of the Netherlands a proposal for finding longitude at sea by means of the eclipses of Jupiter's satellites (*Le opere di Galileo Galilei* [Florence, 1890–1909] **16**: pp. 463–469). Nowhere in that proposal does he claim the invention of the telescope for himself, and, in fact, specifically credits the invention to a Dutchman (p. 464).

procul a me semper fuit mala haec voluntas, sed tamen cuique suum retribuere divinum cum sit, authori vero inventum restituere suum gloriamque in animo alte repostum habui.

Galilaeum autem id non invenisse, sed ab Hollandis habuisse, qui ob viciniam Selandiae crediti fuerunt veri Inventores, patet ex gravissimo & probatissimo Authore Italo *Vittorio Siri,* qui sic de Galilaeo est in Mercurii sui historia tom. 2. L. 3. in fine.[50]

Trovandosi in Venetia, riseppe che in Olando erano state ritrovate le lunete, col cui beneficio gli oggetti visibili si rendevano indistanti all'occhio, benche fossero in sito lontano; senza vedere la forma di questo instrumento, si mise à specularne la struttura, e come potesse essere formato, e finalmente gli sorti di rivenire il Telescopio, vulgarmente chimato il canocchiale di Galilei, onde meritò testimonianze d'istimae d'aggradimento della munificenza del senato.

Id est, cum Venetiis esset, audivit conspicilia reperta fuisse in Hollandia, quorum beneficio objecta visibilia ab oculo remota licet revocabantur, formam ejus instrumenti, licet non vidisset speculatione sua ac ratiocinio, eam scrutatus est, adeo ut tandem Telescopium vel Tubum Galilaeum vocatum invenerit, quare Senatus munificentiis merito honoratus fuit.

Laudo equidem eius ingenium, ejusque acumen, sed docto Viro & ad curiosa proclivi, minima porta aperta cum sufficiat ad aliquid detegendum, non dubium est, quin ille relatu figuras vitrorum Telescopii nostri Selandici (pro quo Hollandiam posuerunt tanquam magis notam) acceperit, sic idem captum fuit a multis, fama volante, ut ab a *Porta, Fratre Paulo* &c. sed semper inventi radix artifici Middelburgensi debetur, reliquis vero rerum ejus ope detectarum praecipua gloria, sed non tota, cum artifex noster non penitus ignarus, multa etiam observasset, quae alii sibi etiam arrogarunt.

Cap. X.

Metium Hollandum Telescopium non invenisse, neque Cornelium Drebellium.

Exclusis Gallis, Hispanis, Italis, &c. Hollandi tantum nobis expellendi supersunt, apud quos Cornelius Drebel & Metius Hollandi, Alckmaërenses de invento non invento ab iis contendunt, certissimum enim habeo testimonium eos convenisse opificem nostrum Middelburgensem & ab eo rem totam accepisle [*sic*]. Drebel enim & Metius audita historia Viri Middelburgensis, qui Telescopium invenit, & Domino Mauritio Principi, Archiducique Alberto Telescopia dederat, Alckmaëre urbe relicta se Middelburgum contulerunt, ut Virum nostrum convenirent, & ab ore proprio rem ediscerent.

Male ergo Cartesius Alckmaërensem Metium Telescopii Inventorem facit in sua Dioptrica, sed rem sic vulgo creditam retulit, ut ferri audiverat.

[50] *Del Mercurio overo historia de' correnti tempi di Vittorio Siri* 2 (Geneva, 1649): pp. 1720–1722.

Cap. XI.

Telescopium Middelburgensi Artifici deberi.

Patet ergo Hollandos etiam Telescopium non reperisse, *Metio Drebellioque,* reliquisque expellendis ab hac victoria expulsis gloria tota Middelburgensi Civi remanet, de qua re confirmativum audiamus *Hieronymi Sirturi* testimonium lib. de Telesc.[51] p. 1, c. 1. qui rem satis recte enodavit, licet quaedam alio modo quam revera sunt, contorserit, relatione non satis exquisite audita.

[Borel here quotes the passage "Prodiit anno 1609 . . . examinavi & similia confecti," see p. 48, above.]

Hac citatione patet, omnes nationes sibi hoc inventum tribuisse, patet etiam Civem nostrum Middelburgensem primum fecisse Telescopium, nam quae de ignoto genio refert, sunt vulgi somnia, & nuntia e longinquo contorsa. Nomen etiam Artificis aliquomodo immutat, sed parum ut infra dicetur. Rem autem a nullo peregrino accepit, sed ipse, curiositatis ergo, multa conspicilia probans, vel quia opticam amabat, aliquid detegere cupiens, istud foelicissime invenit, sed ob suae tenuitatem fortunae, rem ignoto tribuere maluit, vel Principum jussu secretum tacuit, ut soli eo uterentur adversus hostes, & in patriae gratiam proprio damno, & gloriae suae privatione reticuit.

Historia etiam genii seu peregrini vera etiam est, sed non de isto, at de alio ejusdem Urbis spicillorum artifici, ut infra dicetur, & sic Artifex noster a genio secretum non habuit, nec solus erat spicillorum artifex suae Urbis, ut Sirturus dixit. Peregrinus enim ille Genius vocatus, se Middelburgum contulerat, subolfacto invento nostri veri Inventoris, & cum se ad alium, loco ejus, forte contulisset, alter ab eo rem ingeniose etiam accepit, quod forsan aliquid jam prius confuse de vicini sui fama audivisset.

Cap. XII.

De inventoris vero Nomine.

Zacharias Joannides, Inventor est versus Telescopii; eratque autem Conspiciliorum Artifex peritissimus, Middelburgensis Zeelandus, qui anno 1590. admotis (non fato quodam) oculo duobus conspiciliis, nempe lentem cavam & convexam, Tuboque immissis felicissime (ut vult Cartesius) invenit Telescopium. Sed rerum abstrusarum & reconditarum in Optica, quae callebat, desiderio flagrans, ad haec tentanda motus fuit: quare male conqueritur Cartesius, hoc inventum adeo utile & mirandum, scientiarum nostrarum opprobrio, vagis experimentis, & casui fortuito deberi. Telescopium ergo Artifex nostet [*sic*] rimando ex professo indagavit, & tubos 16. pollicum primo fecit, optimos tamen, quem Principibus Mauritio, & Archiduci Al-

[51] See pp. 48, 50, above.

berto, ut testimoniis infra probabimus, obtulit, pro quibus pecunias accepit, rogatus ne rem amplius propalaret, ut ipsi eo uti interim ad bellica possent, quibus ille in patriae gratiam obtemperavit, & sic diu delituit in obscuro Inventor noster.

Invenit praeterea Microscopium ut testimoniis patebit sequentibus.

Novus noster Dedalus Dioptricae non ignarus & ratiocinio eximio pollens, statim ad astra detegenda, aliaque nova se accinxit, septem in Ursa insignes novas stellas detegit, ut infra videre est, &c. Dedalus, inquam, hic novus, absque alis coelum petens, plus uno tubo oculoque, quam Argus vel Lynceus vidit, nec astra recondita oculum ejus effugerunt. In Luna etiam maculas primus observavit, & deinde Galilaeus ejus exemplo eadem etiam observavit exactius. De eo optimo jure, quae de Magno Hevelio dicta fuere, dici possunt:

Scilicet audaci speculo scrutatus Olympum,
Et per vagatus Astra suprema gradu,
Hactenus invisos oculis nunc subjicit Orbes,
Inque Luna vastos Regionum cernere tractus Facit.[52]

Sed de hisce fuse egimus Libro nostro de Terrestrium Globorum pluralitate, in quo Lunam & reliqua Astra Mundos esse, Terram vero Stellam probavimus.[53]

Eum profecto laudibus omnes Musarum Alumni extollere, & Urbem tanti Inventoris alumnam celebrare debent. Hunc ergo ut spero

——— *neque ventura silebunt*
Lustra, nec obscuro rapiet sub nube Vetustas.

Eumque inter Beatos illos reponent Authores, quos aequus Jupiter amavit, & inter gloriosos illos novarum in coelo Observationum Observatores tanquam fulgentissimum Astrum reponent, cum primum gradum aliis, ac iter proposuerit certissimum, januasque aperuerit.

Foelices animae quibus isthaec scandere primum,
Inque domos superas scandere cura fuit.

Nec gloria sua privandus est Iohannes, ejus Filius, qui sedulo Arti huic perficiendae cum Patre incubuit: Nec etiam Hans la preii, Lippersein a Sirturo vocatus, ejusdem Urbis Middelburgensis cum reliquis Civis, qui idem Inventum casu accepit, & fere sponte post minimam cognitionem perfecit, ut infra dicetur.

Cap. XIII.

De iis, quae Joannes Zachariae, Joannidis Filius, Invento paterno in Coelo detexit.

Cum audiverim quaedam a Filio Inventoris, non contemnenda, in coelo detecta fuisse, in laudem Ejus, Patriaeque suae, ea publica facere volui, quare accipe Lector, quae ipse Epistolis suis communicarit, licet adhuc ea mihi comprobare non licuerit. Observavit autem globulum lucidum quasi Lunam alteram in Luna, cujus radiis instar melopeponis dividitur; ut & septem Stellas novas in Ursa majori, quas nomine septem Unitarum Belgii Provinciarum, sub Sagittarum fasciculi specie, insigniendas esse existimavi, aliorum exemplo, qui Borboniorum,[54] Medicaeorum,[55] & Urbanoctaviorum [56] nomen Astris a se detectis donaverant, ut, astris circum-solaribus & circum-jovialibus, a Galilaeo & Reyta.

Cap. XIV

Testimonia egregia, pro Inventoribus supradictis, quibus ea, quae a nobis dicta fuerunt, comprobantur.

Jam tempus adest testimonia exhibendi, quibus optime convincitur Middelburgensi Civi Telescopii & Microscopii Inventum deberi: En illa igitur ex Autographis.

[The first two documents are the depositions taken in Middelburg in favor of Lipperhey and Janssen, see pp. 56–58, above.]

GUILLELMIUS BORELIUS
Belgii Uniti Legatus,
PETRO BORELLO Medico Regio
S.P.

Petis a me, ut quae comperta habeam de Telescopii syderei inventione, tibi per epistolam, id est, breviter, declarem. Accipe igitur quae dicam. Middelburgum

[52] The original lines in the laudatory poem by Vincentius Fabricius, immediately following the *Ad Lectorem* of Hevelius's *Selenographia* (Gdansk, 1647), read:
Scilicet audaci speculo scrutatus Olympum,
Et pervagatus astra sublimi gradu,
Hactenus invisos, oculis nunc subjicit Orbes.
Vasti Universi proferens pomaeria

[53] *Discours nouveau prouvant la pluralite des mondes, que les astres sont des terres habitees et la terre une estoile, qu'elle est hors du centre du monde dans le troisiesme ciel et se tourne devant le soleil qui est fixe, et autres choses tres curieuses par* Pierre Borel (Geneva, 1657). See Marie-Rose Carré, "A Man between Two Worlds: Pierre Borel and his *Discours nouveau prouvant la pluralite des mondes* of 1657," *Isis* **65** (1974): pp. 322–335.

[54] Jean Tarde argued that sunspots were satellites of the sun, and called these satellites the Bourbon Stars, in honor of the French king, Louis XIII. He put forward this theory (already proposed by Christoph Scheiner earlier) in his *Borbonia sydera, id est planetae qui Solis lumina circumvolitant motu proprio ac regulari, falso hastenus ab helioscopis maculae Solis nuncupati* (Paris, 1620).

[55] Galileo, of course, named the four satellites of Jupiter discovered by him in 1610 after the Medici family.

[56] In his *Novem stellae circa Jovem visae, circa Saturnum sex, circa Martem nonnullae* (Louvain, 1643), Antonius Maria Schyrlaeus de Rheita claimed, among other things, to have detected five new satellites of Jupiter, which he named after Pope Urban VIII. Actually, the four satellites discovered by Galileo remained the only ones known until the last decade of the nineteenth century.

Selandorum Metropolis mihi Patria est: juxta aedes ubi natus sum in Foro Olitorio, Templum novum est cujus parietibus nectuntur aediculae quaedum satis humilis: harum unam prope Portam Monetariam Occidentalem inhabitabat Anno 1591. (cum natus sum) quidam conspiciliorum confector nomine Hans, Uxor ejus Maria, qui Filium habuit praeter Filias duas, Zachariae nomine, quem novi familiarissime, quia puero mihi vicino vicinus ab ineunte tenerrima aetate colludens semper adfuit, egoque puer in Officina ipsi saepiuscule adfui. Hic Hans, id est, Iohannes, cum Filio suo Zacharia, ut saepe audivi, Microscopia primi invenere, quae Principi Mauritio Gubernatori & summo Duci Exercitus Belgicae foederatae obtulerunt, & honorario aliquo donati sunt. Simile Microscopium postea ab ipsis oblatum fuit Alberto Archiduci Austrico, Belgicae Regiae Supremo Gubernatori. Cum in Anglia Anno 1619. Legatus essem, Cornelius Drebelius Alckmarianus Hollandus, Vir multorum Secretorum Naturae conscius, ibique Regi Iacobo in Mathematicis inserviens, & mihi familiaris, ostendit illud ipsum instrumentum mihi, quod Archidux ipsi Drebellio dono dederat, videlicet Microscopium Zachariae istius, nec erat (ut nunc talia monstrantur) curto tubo, sed fere ad sesquipedem longo, cui tubus ipse erat ex aere inaurato, latitudinis duorum digitorum in diametro, insidens tribus delphinis ex aere, itidem subnixis, in basis disco ex ligno Ebeno, qui discus continebat impositas quisquilias, aut minuta quaeque, quas desuper inspectabamus forma ampliata ad miraculum fere maxima. Ast longe post, nempe anno 1610. inquirendo paulatim etiam ab illis inventa sunt Middelburgi Telescopia longa syderea, de quibus tibi res est, & unde Lunam & reliquos Planetas, stellas & sydera inspectamus, quorum specimen unum Principi Mauritio etiam obtulit, qui illud inter secreta custodivit, usui futurum forte, in Expeditionibus bellicis. Ut tamen rumor tam mirandi novi inventi increbuit, & jam in Hollandia & alibi de authore loquerentur homines curiosi, Vir quidam hactenus ignotus, ex Hollandia Middelburgum venit apud authorem, inquisiturus super secreto isto, qui cum quaereret conspiciliorum Confectorem in dicta civitate degentem, in aedibus parvis innixis templo novo, casu incidit in Joannem Lapreyum etiam Conspicillificem in vico Caponario etiam aediculas templo novo innitentes inhabitantem; credens esse se apud verum Inventorem, qui exigua tantum distantia ab illo Lapreyo, in altero latere templi dicti & angulo satis obscuro morabatur. Et cum Lapreyo sermones de secreto Telescopii habuit. Qui homo ingeniosus & observator anxius omnium quae vir ille aperuit, etiam quaestiones & lunularum sive lentium comparationes jam longas, jam proximas considerans, post dictum Zachariam Joannidem, egregia industria ac cura eadem Telescopia longa invenit, & confecit ad placitum istius viri peregrini. Quare merito hic Ioannes Lapreius, etiam pro Inventore secundo audiri potest, cum ingenii sui acumine rem non monstratam detexit ex eventu quod dixi, fecitque illa Telescopia sua publici juris, & primus divulgavit.

Res & error tamen brevi sese manifestavit, nam Adrianus Metius Alcmarianus Mathematices Professor, & post eum Cornelius Drebellius supra nominatus, re cognita Anno 1620. Middelburgum venerunt, & non Ioannem Lapreyum, sed Zachariam Ioannidem adierunt, a quo singuli Telescopia pretio compararunt, & multis observationibus & curis, sicut & Galilaeus a Galilaeis Florentinus Italus, & alii multi doctissimi viri rem inventam magnopere illustrarunt, inventi primi tamen honore apud illos duos Middelburgenses in solidum manente. Quibus ego seu primis Middelburgensibus, seu adornatoribus, per hanc meam Epistolam nihil quicquam detractum iri volo. Vale Vir Doctissime, & iis quae experientia & memoria satis certa mihi dictavit, utere si lubet. Dabantur Lutetiis nona die mensis Julii Anno 1655.

[The remainder of this chapter deals with the celestial observations of Johannes Sachariassen.]

Chapter VIII

All nations have claimed the invention of the telescope for themselves

There was no nation which has not claimed for itself the remarkable invention of the telescope: indeed, the French, Spanish, English, Italians, and Hollanders have maintained that they did this, as is evident from Sirtori's book on the telescope, chapter 1, part 1.[41] Nevertheless, it was discovered by none of them, but in fact was discovered in Zeeland in the Netherlands, as will be shown in the appropriate place.

Some suppose that the thing was known to the Ancients, but was kept a secret: however, if it was known to anyone, it was known to Democritus (that great philosopher whose very worthy philosophy I reconcile in my work)[42] who first said that the Milky Way is a collection of stars, and other remarkable things, which were difficult, nay impossible, to explore with the acuity of the naked eye. But his invention lay buried in oblivion until our time.

Others maintain that the thing was known to [Roger] Bacon the Englishman; others attribute it to Battista Porta, who appears to have said something about this matter, although obscurely; nor is there a lack of those who concede it to a certain man from Sedan, named Crepi,[43] an excellent craftsman. But as nothing was published by any one of these, we may rightfully doubt their claims. There was a certain Brother Paul,[44] an Italian, a man of very sharp wit, who knew of the making of lenses, but he had learned it from hearsay, and everything originated in Zeeland, as will be shown more clearly than the light of day.

But there was a great body of opinion for Drebbel,[45] Galileo, and Metius, to each of whom this invention is attributed by many people, and who do not blush to claim the same for themselves, while it is abundantly evident to everybody, indeed, from public testimony that they either met an artisan from Middelburg or borrowed it from him in another way. The majority of serious authors deny this invention to the French, Italians, &c., and concede it to the Hollanders, but still they do not know the matter completely. However, they come closer to the truth, for they have figured out that the thing was invented around this region. But they would rather attribute it to famous mathematicians than to a common artisan.

Thus Descartes has claimed in his *Dioptrique* that the inventor was a Hollander, and seems to think it was Metius.

Nor has there been a lack of those who have dared to report that the thing has come from Spain.[46] But since these reports are no less obscure than the [reports] about the man from Sedan,[47] and since they are approved by very few, we shall not inquire into them further.

Chapter IX

Telescopes were invented not by Galileo but by men from Zeeland

Among this crowd of inventors of telescopes the first to come forward is Galileo, who attributed the invention to himself,[48] and has heretofore been elevated to glory among many nations as the true inventor. And he has heaped his own praises on himself (as is clearly shown in his request presented to the Dutch Republic),[49] for which reason they are commonly called Galilean telescopes. But if any fame is owed to him in the improvement and development of the invention, he should not be deprived of that glory in my opinion, for such ill will has always been far from my purpose. Rather, since it is a heavenly thing to restore to each his own, I have had the deep desire to restore to the true author his invention and his fame.

Moreover, that Galileo did not invent it, but obtained it from Hollanders who, because of their nearness to Zeeland, were thought to be the true inventors, is evident from the very serious and highly respected Italian author Vittorio Siri, who writes thus about Galileo in his *Mercurio,* vol. 2, Book 3, at the end:[50]

[See the Italian passage on pp. 59, above]

That is, "when he was in Venice, he heard that glasses had been invented in the Netherlands with the help of which visible objects were brought close to the eye, even though they were far off. Without having seen the form of this instrument, he set about investigating it by speculation and ratiocination, until at length he devised the telescope, also called the Galilean tube, for which he was properly honored with presents by the senate."

I do indeed praise his ingenuity and his cleverness, but since for a learned man inclined towards curious things, opening the smallest door is enough for the discovery of something, there is no doubt that he received by report the shapes of the glasses of our Zeeland telescope, which they attributed to Holland because it is better known. Thus, as the report circulated, it was seized on by many, such as Porta, Brother Paul, etc. But the origin of the invention is always due to the artisan of Middelburg, while the others have the particular fame of discovering things with his help. But they are not [entitled] to all the fame, since our artisan, who was not entirely ignorant, had made many observations which others also claimed for themselves.

Chapter X

Metius of Holland did not invent the telescope, nor did Cornelis Drebbel

With the French, Spanish, Italians, etc., excluded, it remains for us to eliminate only the Hollanders, among whom Cornelis Drebbel and Metius of Alkmaar claim the invention which was not invented by them. I have, in fact, very certain evidence that they met our artisan of Middelburg, and received the whole thing from him. For Drebbel and Metius had heard the story about the man from Middelburg who had invented the telescope and had given it to Prince Maurice and Archduke Albert. They left Alkmaar and went to Middelburg, in order to visit our man and to learn about the thing from his own mouth.

And so Descartes wrongly makes Metius of Alkmaar the inventor of the telescope in his *Dioptrique,* but he reported the matter thus because it was generally believed, and he had heard it like that.

Chapter XI

The telescope is due to an artisan from Middelburg

It is evident, therefore, that the Hollanders did not invent the telescope, and that with Metius, Drebbel, and the others dismissed from this triumph the entire glory remains to the citizen of Middelburg, about whom we can hear the confirming evidence of Girolamo Sirtori, in his book on the telescope, part I, chapter 1,[51] where he has explained the matter quite correctly (although he has twisted some things so that they differ from the truth), in an account which has not been sufficiently carefully attended to:

"In the year 1609 there appeared a genius or some other man, as yet unknown, of the race of Hollanders, who, in Middelburg in Zeeland, visited Johannes Lippersein, a man distinguished from others by his remarkable appearance, and a spectacle-maker. There is no other [spectacle-maker] in that city, and he ordered many lenses to be made, concave as well as convex. On the agreed day he returned, eager for the finished work, and as soon as he had them before him, raising two of them up, namely a concave and a convex one, he put the one and the other before his eye and slowly moved them to and fro, either to test the gathering point or the workmanship, and after that he left, having paid the [spectacle-]maker. The artisan, by no means devoid of ingenuity, began to do the same, and quickly his wit suggested that these lenses should be joined together in a tube." Etc.

From this citation it is evident that all nations have attributed this invention to themselves, and it is also evident that our citizen of Middelburg first made the first telescope, for what Sirtori reports about the unknown genius is idle fancy and news distorted at a distance. Also, he changes the name of the artisan somewhat, but not much, as will be stated below. However, [the citizen of Middelburg] did not learn this from any stranger, but rather he himself most fortunately invented it by his curiosity, through examining many lenses, perhaps because he liked optics and was eager to discover something. But because of his poverty he preferred to assign the thing to an unknown person, or by order of the princes was silent about the secret, so that they alone might use it against their enemies. And by keeping silent he hurt himself and lost his fame, for the sake of his country.

The story about the genius or stranger is also true, but not about that man, but about another spectacle-maker of the same city, as will be related below. And thus our artisan did not get the secret from the genius, nor was he the only spectacle-maker in his city, as Sirtori said. For that stranger, called a genius, betook himself to Middelburg, having found out about the invention of our true inventor, and when by chance he came to the other spectacle-maker instead of to the inventor, the other one ingeniously also got the same thing from him, about which he had perhaps already heard something confusedly through rumors concerning his neighbor.

Chapter XII

The true name of the inventor

Sacharias Janssen is the true inventor of the telescope, and he was, moreover, a very skilled spectacle-maker, a Zeelander of Middelburg, who in the year 1590 having applied to his eye two glasses (but not by some trick of fate), that is, a concave and a convex lens, and having put them into a tube most fortunately (as Descartes thinks), invented the telescope. But, in fact, he was urged to try this by his burning desire for abstruse and deep secrets in optics, which was his profession. For this reason Descartes wrongly complains that this extremely useful and wonderful invention is due to trial and error and a fortuitous event, to the disgrace of our sciences. Our artisan, therefore, devised the telescope by deliberate investigation, and first made tubes of 16 inches, and gave the best to Prince Maurice and Archduke Albert, as we shall show below in the testimonies, for which he received money and was asked not to divulge the thing further, so that they themselves could in the meantime use it in military affairs, with which he complied for the sake of his country. And so our inventor was hidden in obscurity for a long time.

Besides this, he invented the microscope, as will become evident from the testimonies below. Our new Daedalus of Dioptrics was not an ignorant man and, being possessed of a good intellect, he at once undertook to examine the stars and other new things, and detected seven conspicuous new stars in the Bear, as will be seen below, etc. This new Daedelus, I say, searching the sky without wings, saw more with one tube and one eye than Argus or Lynceus, nor did hidden stars escape his eye. He also first observed spots on the Moon, and afterwards, after his example, Galileo also observed the same, more exactly. What was said of the great Hevelius can with the greatest justice be said of him:

With a bold looking glass he scrutinized Olympus
And passing through the highest rank of stars
He now conquers hitherto unseen orbs
And in the Moon he makes us see vast tracts of regions.[52]

But of these things we have spoken amply in our book on the plurality of terrestrial globes, in which we have proved that the Moon and the other stars are worlds, and that the Earth is indeed a star.[53]

Surely, all followers of the Muses ought to extol him with praises and celebrate the city which nourished so great an inventor. Therefore, as I hope,

The years to come will not be silent about him
Nor will time drag him under a cloud of obscurity

And the authors may include him among those blessed men whom fair-minded Jove loved, and number him as a very bright star, so to speak, among those glorious observers of new things in the heavens, since he took the first step and pointed out a very definite path for others, and opened the door.

Happy souls whose task it was to ascend thither first
And climb into the realms above

Nor is Johannes, his son, to be deprived of his glory, who with his father took up the burden of zealously perfecting this art. Nor also Hans Laprey, called Lippersein by Sirtori, like the others, a citizen of the same city of Middelburg, who received the same invention by chance and perfected it almost by himself, after the shortest acquaintance with it, as will be explained below.

Chapter XIII

What Johannes Sachariassen, son of Janssen, detected in the heavens with his father's invention

When I heard that some things not to be despised had been discovered in the heavens by the son of the inventor, I wished to make them public in honor of him and his country, for which reason, dear reader, read what he himself has communicated in his letters, although I have not been able to confirm them as yet. He observed a shining globe, like another moon, within the Moon, by whose rays [the Moon] is divided like a melon; and also seven new stars in the Great Bear, which I thought should be distinguished by the name of the seven United Dutch Provinces, like a bundle of arrows, after the example of others who had given the name of the Bourbons,[54] the Medicis,[55] and Urban VIII[56] to stars discovered by themselves, that is, the circumsolar and circumjovial stars of Galileo and Rheita.

Chapter XIV

Distinguished testimonies in favor of the above-mentioned inventors, and in confirmation of what we have said

Now the time has come to present the testimonies by which one may be quite convinced that the invention of the telescope and microscope is owed to a citizen of Middelburg. Here they are, therefore, from the autographs:

[The testimonies in favor of Janssen and Lipperhey given here are presented above, pp. 56–58.]

WILLEM BOREEL
Ambassador of the United Netherlands
Greets PIERRE BOREL, Royal Physician

You ask me to tell you in a letter, that is to say briefly, what is known to me about the invention of the sidereal telescope. Receive therefore what I shall tell you. The city of Middelburg, the capital of Zeeland, is my native city. Near the house where I was born, in the vegetable market, stands the new church, attached to the walls of which are some rather humble little houses in one of which, near the western door of the Mint, lived in the year 1591 (when I was born) a certain spectacle-maker by the name of Hans with his wife Maria, who had, besides two daughters, a son named Sacharias whom I knew intimately because, being a neighbor, he was always near, and we played together from an early age. And as a boy I was very often in the workshop with him. Here Hans, or Johannes, with his son Sacharias, as I have often heard, first invented the microscope, which they presented to the stadholder Prince Maurice, the commander in chief of the army of the Dutch Federation, and they were given some reward. Afterwards a similar microscope was presented by them to the Archduke Albert of Austria, the supreme Governor of Royal Belgium. When I was sent to England as ambassador in 1619, Cornelis Drebbel, a Hollander from Alkmaar, a man learned in many secrets of nature, and there serving King James as mathematician, and an intimate friend of mine, showed me that very same instrument which the archduke had given to that same Drebbel, that is to say, the microscope of that Sacharias. And it was not with a short tube, such as these instruments have now, but about a foot and a half long, the tube being of gilded brass, two inches in diameter, resting on three dolphins made of brass, which were supported in turn by the disc of the base made out of ebony. This disc contained some small things or sweepings laid in, and when we looked down on these [through the instrument] they appeared almost miraculously amplified. Moreover, long afterwards, that is, in the year 1610, when they pursued their investigations little by little, long sidereal telescopes were also invented by them in Middelburg, in which you are interested, and with which we inspect the Moon, the other planets, the stars, and constellations. One example of these [instruments] was also presented to Prince Maurice, who kept it a secret, perhaps for future use in military expeditions. As, however, the news of such an extraordinary new invention spread, and already in Holland and elsewhere inquisitive people were talking about the inventor, a certain man, as yet unknown, came from Holland to Middelburg to the inventor's house, making inquiries about that secret. When he asked after the spectacle-maker living in the said city, in the little houses leaning against the new church, he came by chance upon Hans Laprey, also a spectacle-maker and also living in the Capoen Street in the little houses next to the new church. And he believed himself to be at the house of the true inventor, who lived only a short distance from this Laprey, on the other side of the said church in a rather hidden corner. And he talked about the secret of the telescope with Laprey who, being a clever and observant person, attentive to everything the man revealed to him, and also considering his questions and the comparisons of the lunulae or lenses, both long and short, invented, by uncommon labor and zeal, the same long telescopes after the said Sacharias Janssen, and made them to the satisfaction of the stranger. Therefore this Hans Laprey can justly be regarded as the second inventor, since by the sharpness of his wit he worked out a thing which was not shown to him, with the result which I have related, and first made those telescopes of his public.

Both the instrument and the mistake, however, shortly became known. For Adriaen Metius of Alkmaar, professor of mathematics, and after him Cornelis Drebbel, mentioned above, came to Middelburg in 1620, when the matter was well known, and did not go to Hans Laprey, but to Sacharias Janssen, from whom each obtained a telescope at a price, and by many observations and efforts, like Galileo Galilei the Italian from Florence, and many other very learned men, they made the invention very famous, with the honor of the first invention, however, remaining firmly with those two Middelburgers. I wish nothing whatever to be detracted from the first men of Middelburg, or from the embellishers, by this letter of mine. Farewell most learned Sir, and use as you please those things which experience and a pretty reliable memory have given me to know. Paris, the ninth day of July, 1655.

VII. INDEX